T0226103

Alternative Fuels in Ship Power Plants

Xinglin Yang · Zongming Yang · Huabing Wen ·
Viktor Gorbov · Vira Mitienkova · Serhiy Serbin

Alternative Fuels in Ship Power Plants

Application of Alternative Fuels

Recommended by the Ministry of education and science
of Ukraine as a study guide

Xinglin Yang
Jiangsu University of Science
and Technology
Zhenjiang, Jiangsu, China

Zongming Yang
Jiangsu University of Science
and Technology
Zhenjiang, Jiangsu, China

Huabing Wen
Jiangsu University of Science
and Technology
Zhenjiang, Jiangsu, China

Viktor Gorbov
Admiral Makarov National University
of Shipbuilding
Mykolayiv, Ukraine

Vira Mitienkova
Admiral Makarov National University
of Shipbuilding
Mykolayiv, Ukraine

Serhiy Serbin
Admiral Makarov National University
of Shipbuilding
Mykolayiv, Ukraine

ISBN 978-981-33-4852-3   ISBN 978-981-33-4850-9 (eBook)
https://doi.org/10.1007/978-981-33-4850-9

Jointly published with Shanghai Scientific and Technical Publishers, China
The print edition is not for sale in China Mainland. Customers from China Mainland please order the print book from: Shanghai Scientific and Technical Publishers.

This Springer imprint is published by the registered company Springer Nature Singapore Pte Ltd.
The registered company address is: 152 Beach Road, #21-01/04 Gateway East, Singapore 189721, Singapore

About This Book

The publication presents data on the impact of ship power plants on the atmosphere and methods for reducing the environmental load. The main characteristics of alternative fuels which have a great potential for use in transport and stationary power engineering are considered, as well as methods for their obtaining. The experience, prospects and forecasts of application of alternative fuels on ships are analyzed. The issues of increasing the efficiency of the use of liquefied natural gas and biodiesel fuels in ship power plants are considered by determining reasonable thermohydrodynamic parameters and mass-dimensional characteristics of fuel systems.

Introduction

The monograph is intended for scientific personnel, engineers, and postgraduate students who specialize in the field of alternative fuels application in ship and stationary power plants. It can be helpful for students of the specialties "Ship Power Plants and Equipment," "Heat Power Engineering" and related specialties of higher technical educational institutions.

Over the past two decades, the scale of application of alternative fuels in various means of transport and in stationary power engineering has increased substantially. This is caused by the continuous strengthening of legislative requirements for emissions and the desire of many governments to reduce the dependence on fossil fuel export. Most ship owners are motivated to switch to alternative fuels because they are not willing to pay high fines for emissions of nitrogen oxides, sulfur and carbon, which are necessarily present in the exhaust gases of heat engines and boilers operating on oil fuels. Other economic reasons include reduction of capital costs (this is a low-cost technology for emissions decrease) and operating costs (the price for individual alternative fuels is already lower than that for oil).

Currently, a fairly large number of alternative fuel types raises questions about the opportunity and feasibility of their application on various transport vehicles, in particular, in water transport. Given its specificity, a relatively small number of unconventional fuels can be used on ships, and it is necessary to formulate the main problems and issues that need to be addressed for their use. Analysis of the experience available on the use of alternative fuels in water transport has shown that operation of ship power plants on water fuel emulsions, liquefied natural and oil gas, biodiesel fuels and their mixtures can be considered successful. The use of other types of fuels is still episodic and does not suggest their widespread application on ships any time soon.

The main advantages of the above-mentioned fuels are as follows: They are produced or extracted on an industrially significant scale, can be used in ship engines and boilers virtually without modifying the design and operating characteristics of the latter and do not pose a threat to human life and health. In addition, ecological characteristics of ship power plants are being improved, and the price of alternative fuels is comparable to that of oil fuels. All of this leads to a relatively wide use of alternative fuels on ships.

Introduction of the new types of fuels with properties different from standard fuels in the sea fleet necessitates modification of the design of fuel systems and sometimes even of the ship power plant itself, which may lead to a change in a number of the SPP indicators. Therefore, it is important to choose the rational parameters of the fuel system equipment and the working media already at the stage of conceptual design, so that they would ensure optimal characteristics of the system and its normal operation. There is a vast experience in the creation and operation of ship fuel systems running on oil fuels; their principal schemes have been worked out, and in many cases, the rational parameters of working media and equipment have already been determined. Meanwhile, these data are very limited or completely absent for alternative fuels. Solution of these problems constitutes the subject matter of this monograph. The authors have developed mathematical models and methods for calculating the parameters of fuel systems for biodiesel fuels and liquefied natural gas. Recommendations for choosing the rational parameters of these systems are given, as are schematic solutions of the fuel systems, recommendations for selecting equipment, storing and preparing the fuels.

Application of the materials described in the monograph will provide the SPP designers with a reliable tool for choosing rational characteristics of the fuel systems operating on alternative fuels and improving the efficiency of their application on ships.

Contents

Chapter 1
Substantiation of Alternative Fuels Utilization

1.1 Contribution of Marine Transport to the Global Atmosphere Pollution

Air pollution caused by the pollution emissions from ships is a serious environmental problem, especially in the areas of equatorial waters and inland waterways. Currently more than 96% of vessels are equipped with diesel ship power plants as the propulsion engines, so the constituents formed in the process of fuel oils burning can contaminate the environment (Table 1.1) [1]. Accordingly, the above-mentioned constituents of the heat engines and boilers exhaust gases have a negative impact on the environment and human health.

Nitrogen oxides of the internal combustion engines' exhaust gases consist of 80–90% of NO and 10–20% of NO_2 [2, 3]. Nitric oxide is toxic. In case of its inhalation it affects the respiratory tract. Nitrogen dioxide is very toxic, it also affects the respiratory tract and in high concentrations it may cause pulmonary edema [3].

The fuel compounds containing sulphur regardless of their chemical structure under their combustion within the engine are converted mainly to sulphur dioxide. The content of the sulphur dioxide in the exhaust gases depends on the sulphur compounds concentration in the fuel. SO_2 is toxic for humans.

It causes asphyxiation and pulmonary edema in case of high concentrations. It can be dissolved in water to form sulphuric acid [3].

Among the hydrocarbons (HC) contained in the exhaust gases of diesel engines, the most carcinogenic is polycyclic hydrocarbon benzopyrene, which can cause malignant tumors of animals and humans [2]. Methane (CH_4) comprises 2 to 6% of the total amount of hydrocarbons in the engine exhaust gases. It is formed under the thermal decomposition of hydrocarbon fuel compounds [3]. Methane greenhouse capacity is 21 units [4]. Hydrocarbons can form ozone when combined with nitrogen oxides, and ozone is also referred to as greenhouse gases [5].

Carbon monoxide (CO) is formed during the carbon-containing materials combustion in terms of insufficient amount of air for complete combustion and formation of carbon dioxide as the final combustion product. CO is strong poison for people,

© Shanghai Scientific and Technical Publishers 2021
X. Yang et al., *Alternative Fuels in Ship Power Plants*,
https://doi.org/10.1007/978-981-33-4850-9_1

Table 1.1 Pollutants in diesel exhaust

Solids	Liquids	Gases
Soot – Primary particles – Agglomerated particles – Sulphates Ash – Oil additives Engine wear particles Inorganic fuel and air contaminants	Soluble organic fraction (SOF) – Fuel derived – Oil derived Poly nuclear aromatic Hydrocarbons (PAH) Sulphuric acid	Nitric oxide (NO) Nitrogen dioxide (NO_2) Unburned hydrocarbons (HC) Carbon monoxide (CO) Carbon dioxide (CO_2)

because this substance can connected with the blood hemoglobin in 200–300 times faster than oxygen to form a stable complex compound of light-red color called carboxyhemoglobin. Thus the processes of oxygen distribution and cellular respiration are blocked which leads to oxygen distribution disorder within the human body, hypoxia [3].

Soot is a particulate matter (PM) of hydrocarbon materials with carbon content up to 99%. The black smoke in the output is the reason of the engine exhaust gases PM. Soot is a mechanical contaminant of human lungs, but it is much more dangerous as an adsorbent and active transporter of carcinogens, particularly benzopyrene [2].

The exhaust gases of internal combustion engines can contain low-molecular aldehydes (organic compounds containing the aldehyde group CHO), formed during the early stages of high-temperature oxidation of hydrocarbon fuels [3]. The aldehydes can accumulate in the human body. In addition to systemic toxicity they also are neurotoxins, some of them are carcinogenic [6].

Carbon dioxide (CO_2) is non-toxic for humans, at the same time it is greenhouse gas, directly impacting global warming regardless of its origin. CO_2 emissions global level is regulated by the Kyoto Protocol [5].

In addition to diesel engines, the gas turbine and steam turbine engines and units are also used. The fuel type and the boiler or engine type affect the ship power plants emissions level. Table 1.2 represents the emissions data for different types of engines when running on heavy fuel oil (HFO) and marine diesel oil (MDO) with an average sulphur content of 3% and 1%, respectively [7]. In accordance with these data, the emissions of nitrogen oxides, carbon monoxide, volatile organic compounds (VOC) significantly increase in terms of diesel engines operation. The steam-turbine units have the highest soot emissions, as the fuel combustion occurs within the boilers. The engine type does not affect the carbon dioxide emissions, and SO_x emissions depends only on the sulphur content of the fuel.

The marine transport emissions of nitrogen and sulphur oxides into the atmosphere are the most serious problem at the moment. Marine vessels emissions are up to 14% of the total fossil fuels emissions and 16% of the total sulphur emissions of oil products burning. On average, it is about 10 million tons of NO_x and 8.5 million tons of SO_x per year. Ship engines exhaust gases harmful substances emissions can make a major contribution to the continental pollution of inland hundreds of kilometers.

Table 1.2 Default basic emission factors (kg/ton of fuel)

Engine types	NO_x	CO	CO_2	VOC	PM	SO_x
Steam turbines–HFO engines	6.98	0.431	3200	0.085	2.50	60
Steam turbines–MDO engines	6.25	0.6	3200	0.5	2.08	20
High speed diesel engines–HFO	70	9	3200	3	1.5	60
High speed diesel engines–MDO	70	9	3200	3	1.5	20
Medium speed diesel engines–HFO	57	7.4	3200	2.4	1.2	60
Medium speed diesel engines–MDO	57	7.4	3200	2.4	1.2	20
Slow speed diesel engines–HFO	87	7.4	3200	2.4	1.2	60
Slow speed diesel engines–MDO	87	7.4	3200	2.4	1.2	20
Gas turbines	16	0.5	3200	0.2	1.1	20

So, for example SO_2 and NO_x and their atmospheric products emissions will resist for 1–3 days and spread to a distance of 400–1200 km. The content of nitrogen oxides in the atmosphere is the main reason for the smog formation, which is a problem of the major port cities. Side by side with this NO_x as SO_x can be washed from the atmosphere by rain and increase the acidity of the soil.

The total contribution of marine transport to the global environmental pollution depends on the amount of fuel consumption, primarily oil, but the available data differ significantly depending on the source of information (Table 1.3) [5, 8–10].

One of the first researches was prepared for the International Maritime Organization (IMO) by three consultants from Norway: Norwegian Marine Technology Research Institute (Marintek), the company Econ Analyse, the classification society Det Norske Veritas (DNV) and the four research teams from Carnegie Mellon University, USA. The report on emissions levels of carbon, nitrogen and sulphur oxides was based on the estimation of marine fuels global consumption and the statistical model. It was found that the marine transport emitted for about 1.8% of CO_2 global emissions in 1996. The research was based on data from the National Oceanic and Atmospheric Administration Management (US-NOAA) and confirmed that about 80% of the marine transport emissions fall on the areas near the world's coastlines [5].

The Endreson et al. (2003) research was prepared with the help of DNV and the University of Oslo. The researchers developed a statistical model of marine fuels consumption and got the results that were comparable with similar studies carried out for the IMO in 2000. This happened due to the fact that the assumptions were chosen to agree the simulation results with the statistics of fuel consumption. In addition to determining the volume of NO_x, SO_x, PM, CO and CO_2 emissions the research also provided the determination of the marine vessels emissions influence on the ozone formation, sulphate deposits and methane concentration in the atmosphere [5].

The research of Corbett and Koehler (2003) and Eyring et al. (2005) was carried out mainly for more accurate assessment of emissions from international marine traffic. Using the methodology of "bottom-up" (bottom-up methodology is applied

Table 1.3 Fuel consumption, emissions and percent contribution to estimated global inventories (in parenthesis) from international ships greater than 100 tons GRT

Source	Inventory years	Year of publication	Fuel consumption (10^6 metric tons)	NO_x, (10^6 metric tons) (%)	SO_x, (10^6 metric tons) (%)	PM_{10}*, (10^6 metric tons)	CO_2, (10^6 metric tons) (%)
Eyring et al. (DLR institute for atmospheric physics, Germany)	2001	2005	280	21.4 (29)	12 (9)	1.7	813 (3)
Corbett and Koehler (college of marine and earth studies, university of delaware, USA)	2001	2003	289	22.6 (31)	13 (9)	1.6	912 (3)
Endreson et al. (DNV)	1996	2003	158	12 (17)	6.8 (5)	0.9	501 (2)
IMO	1996	2000	120–147	10 (14)	5 (4)	–	419 (1.5)

*PM_{10}—soot particulates of the size under 10 μmin diameter

in solving complex problems, when they are divided into many small ones, and each is solved increasingly) Corbett and Koehler estimated the usage of marine fuel in 289 million tons, which is almost twice as much than according to the IMO (2000) and Endreson et al. (2003) data. As a consequence, the data analysis showed that the contribution of maritime transport to the global environmental pollution from all the sources equals to nearly 30% for nitrogen oxides and 9% for sulphur oxides. The results obtained by Eyring et al. were also based on a similar "bottom-up" methodology, corresponding to the one used by Corbett and Koehler. The difference in the results of the research mentioned above occurred due to the fact that they were based on independent reports on marine fuels sales. Furthermore, Eyring et al. report contained the facts indicating that the data on the fuels sales volume were likely to be erroneous. Indeed, comparison of energy statistics with the number of vessels and circulation of goods during a long period of time showed that the international data on the marine fuels sales volume are significantly underestimate the actual consumption in this sector [5].

Further growth in marine freight traffic will lead to an increase of the maritime transport impact on the global environmental pollution. Figures 1.1 and 1.3 show the forecast of world emissions from marine transport up to 2050. The data in Figs. 1.2 and 1.3 show the contribution of international marine traffic to global air pollution according to the forecast until 2050 based on the International Panel of Climate Change (IPCC) [5].

Figure 1.2 does not represent data on the relative soot contamination by water transport comparing to Fig. 1.1, because the assessment of the soot global emissions has not been conducted yet. Figures 1.1 and 1.3 show that under the period from 2010 to 2050 the emissions volume from the international marine traffic almost doubles

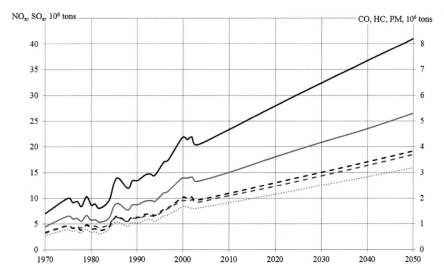

Fig. 1.1 Emissions of NO_x, SO_x, CO, HC and PM from international marine traffic, million tons

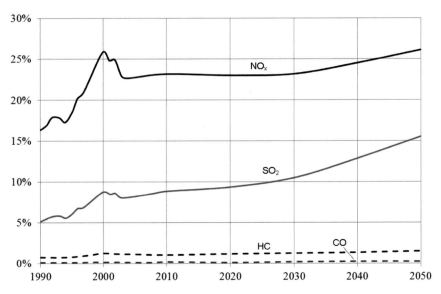

Fig. 1.2 Relative emissions of NO_x, SO_x, CO, HC from international marine traffic in 1990–2050

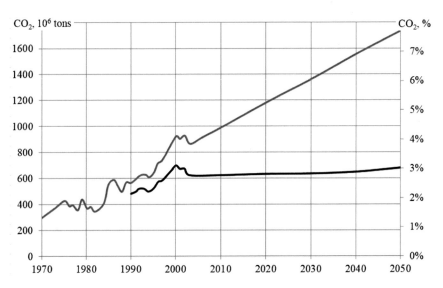

Fig. 1.3 CO_2 emissions from international marine traffic in absolute (million tons) and relative (% of global emissions) values

for NO_x and PM, rises by 80% for SO_x, and about 90% for HC and CO, and 70% for CO_2. At the same time the contribution into global air pollution for vessels does not increase slightly—HC emissions (from 1.3 to 1.7%), CO (from 0.2 to 0.25%) and

CO_2 (from 2.7 to 3%), nearly 4% of increase in NO_x emissions and 8% of sulphur dioxide emissions increase [5].

1.2 Demands of the International Conventions for Preventing Pollution from Marine Vessels

To reduce the negative impact on the atmosphere the number of legislative restrictions on emissions from marine vessels is set. In particular this applies to IMO regulations in the framework of the International Convention for the Prevention of Pollution from Ships (MARPOL 73/78), particularly the Annex VI of MARPOL—"Prevention of Air Pollution from Ships". The instructions in the current Annex set limits on emissions of sulphur and nitrogen oxides from ship power plant and prohibit the uncontrolled emissions of substances that deplete the ozone layer of the atmosphere.

In September 1997, in London, at the diplomatic level the proposals to limit the exhaust gases emissions from the marine diesel engines were developed by IMO. The main guiding document was the Annex VI to MARPOL 73/78, including the "Technical Code on the Control of Nitrogen Oxides Emissions from Ship's Diesel Engines". The "Technical Code" is an international standard that establishes the procedures and rules of ecological certification of marine diesel engines both at the factory and during the operation of the engines in the marine environment.

The regulation 13 of the Annex VI of MARPOL 73/78, including "Technical Code on the Control of Nitrogen Oxides Emissions from Ship's Diesel Engines" is applied to engines with an effective power of more than 130 kW installed on ships constructed after January 2000, except for the engines applied in emergency situations. The engines, which have been significantly reequipped under the timeline also subject to this regulation. The regulation involves prohibiting of the engine operation if the NO_x emissions exceed the limits set by the Tier limiting curves. The requirements of these standards are represented in Table 1.4 [11].

Currently, the local legislative measures are in force: California Regulations (CARB—California Air Resources Board), the corresponding laws of Sweden,

Table 1.4 Requirements of the "technical code on the control of nitrogen oxides emissions from ship's diesel engines"

Different levels (Tiers)	Ship construction date on or after	Total weighted cycle emission limit (g/kWh) n = engine's rated speed (rpm)		
		$n < 130$	$130 \leq n < 2000$	$n \geq 2000$
Tier I	1 January 2000	17.0	$45 \cdot n^{-0,2}$	9.8
Tier II	1 January 2011	14.4	$44 \cdot n^{-0,23}$	7.7
Tier III*	1 January 2016	3.4	$9 \cdot n^{-0,2}$	1.96

*Tier III controls apply only to the specified ships while operating in Emission Control Areas (ECA) established to limit NO_x emissions, outside such areas the Tier II controls apply

Norway and others in addition to the global international restrictions on nitrogen oxides emissions developed by IMO. The requirements of such regulations are rather rigid and apply to ships diesel engines, which operate in the territorial waters of these countries or regions. For example, according to CARB Regulations, the concentration of nitrogen oxides for newly constructed ships should not exceed 130 ppm (0.013%). For comparison it can be noted that typically the concentration of nitrogen oxides, unless special measures are taken, is 1200–1500 ppm. Sweden has introduced a limit on nitrogen oxides emissions for local vessels (ferries)—no more than 2 g/(kWh), which necessitates the use of a catalytic reactor. Norway has set the goal to reduce NO_x emissions from marine local traffic in 10 thousand tons per year (20%). A penalty for nitrogen oxides emissions was introduces in 2007 in the amount of 1 765 EUR/t for all the vessels that were in the territorial waters of Norway [12, 13].

The International Maritime Organization has introduced the control regulations at the global level, limiting the sulphur content in marine fuels. The Annex VI to the MARPOL 73/78 Convention operates in international and territorial waters and prohibits the use of fuel with a sulphur content exceeding 3.5% and by 2020 the allowable sulphur content will have been 0.5%. The Annex VI contains provisions allowing setting the specific control areas of SO_x emissions (SECA—Sulphur Emissions Control Areas) with more strict control of sulphur oxides emissions. Within the SECA areas this value should not exceed 0.1%. The effectiveness of measures taken by the IMO to reduce the sulphur emissions can be estimated from the data given in Table 1.5 [14].

The use of low sulphur fuels with the sulphur content of 0.5% enables reducing the total SO_2 emissions by nearly 15% only in the SECA areas. At the same time, stricter requirements and use of fuels with $S = 0.1\%$ will reduce the sulphur emissions in 16.2%.

According to the European Parliament Directive 2005/33/EU the maximum sulphur content of fuels used by passenger ships on regular sailings, when they are in the territorial waters of the EU, should not exceed 1.5%. From the 1st of January, 2010 the sulphur content in all grades of marine fuels, which is used by vessels at

Table 1.5 Calculated emissions from ships

Calculation assessment, mln tones	Result 2007	Result 2020
Total fuel consumption by ships	369	486
HFO consumption by ships	286	382
Marine distillate consumption by ships	83	104
Total SO_x emission from ships	16.2	22.7
SO_x emission reduced by current SECAs	−0.78	–
SO_x emission reductions for 0.5% S marine distillate global cap	−12.7	−17.8
SO_x emission reductions in multiple SECA environment with a 0.5% marine distillate SECA cap	–	−3.4
SO_x emission reductions in multiple SECA environment with a 0.5% marine distillate SECA cap	–	−3.7

EU ports in inland waters shall not exceed 0.1% [15]. The restrictions were settled for vessels within the Greece and Spain coastal areas on the content of sulphur in the fuel: not more than 0.2% for light oil and not more than 1% for heavy [7].

CARB adopted requirements for fuels used in diesel generators and auxiliary diesel engines of ocean-going vessels (cargo, cruise ships and other large vessels), located in the 24-mile zone of the California coast. According to these requirements the marine fuels with the sulphur content higher than 0.1% may not be used [16].

In 2012 the restrictions for greenhouse gas emissions from marine vessels (primarily CO_2), the so-called CO_2-index for newly constructed vessels, were imposed by the IMO under the Kyoto Protocol [11, 17–22]. IMO has proposed a formula for determining the energy efficiency index for new vessels under construction EEDI (Energy Efficiency Design Index), which takes into account the amount of CO_2 generated during operation. The EEDI determination applying the formula is possible for the following types of vessels [23]: bulk carriers, tankers, gas carriers, LNG carriers, ro-ro cargo ships (vehicle carriers), ro-ro cargo ships, ro-ro passenger ships, general cargo ships, refrigerated cargo carriers, passenger ships and cruise passenger, containerships.

This formula may not be applicable to a ship having diesel-electric propulsion, turbine propulsion or hybrid propulsion system, except for cruise passenger ships and LNG carriers. The attained new ship Energy Efficiency Design Index (EEDI) is a measure of ships' energy efficiency (g/t · nm) and calculated by the formula below [10, 23, 24]:

$$EEDI = \frac{\left(\prod_{j=1}^{M} f_j\right)\left(\sum_{i=1}^{nME} P_{ME(i)} \cdot C_{FME(i)} \cdot SFC_{ME(i)}\right) + \left(P_{AE} \cdot C_{FAE} \cdot SFC_{AE}*\right)}{f_i \cdot f_c \cdot f_l \cdot Capacity \cdot V_{ref} \cdot f_w}$$
$$+ \frac{\left(\left(\prod_{j=1}^{M} f_j \cdot \sum_{i=1}^{nPTI} P_{PTI(i)} - \sum_{i=1}^{neff} f_{eff(i)} \cdot P_{AEeff(i)}\right) C_{FAE} \cdot SFC_{AE}\right) - \left(\sum_{i=1}^{neff} f_{eff(i)} \cdot P_{eff(i)} \cdot C_{FME} \cdot SFC_{ME}\right)}{f_i \cdot f_c \cdot f_l \cdot Capacity \cdot V_{ref} \cdot f_w}$$

*If part of the Normal Maximum Sea Load is provided by shaft generators, SFC_{ME} and C_{FME} may—for that part of the power—be used instead of SFC_{AE} and C_{FAE}.

In this formula SFC is the specific fuel consumption; C_F is a non-dimensional conversion factor between fuel consumption measured in g and CO_2 emission also measured in g based on carbon content (Table 1.6). The subscripts $ME(i)$ and $AE(i)$ refer to the main and auxiliary engine(s) respectively. C_F corresponds to the fuel used when determining SFC listed in the applicable test report included in a Technical File as defined in paragraph 1.3.15 of NO_X Technical Code.

Capacity is defined as follows:

- For bulk carriers, tankers, gas carriers, LNG carriers, ro-ro cargo ships (vehicle carriers), ro-ro cargo ships, ro-ro passenger ships, general cargo ships, refrigerated cargo carrier and combination carriers, deadweight should be used as *capacity*.

Table 1.6 The value of C_F

Type of fuel	Carbon content	C_F, (t-CO$_2$/t-Fuel)
Diesel/Gas oil	0.8744	3.206
Light fuel oil	0.8594	3.151
Heavy fuel oil	0.8493	3.114
Liquefied petroleum gas (LPG)—Propane	0.8182	3.000
Liquefied petroleum gas (LPG)—Propane-Butane	0.8264	3.030
Liquefied natural gas (LNG)	0.7500	2.750
Methanol	0.3750	1.375
Ethanol	0.5217	1.913

- For passenger ships and cruise passenger ships, gross tonnage in accordance with the International Convention of Tonnage Measurement of Ships 1969, annex I, regulation 3, should be used as *capacity*.
- For containerships, 70% of the deadweight (DWT) should be used as *capacity*.

V_{ref} is the ship speed, measured in nautical miles per hour (knot), on deep water.

$P_{ME(i)}$ is 75% of the rated installed power for each main engine (*i*).

$P_{PTO(i)}$ is 75% of the rated electrical output power of each shaft generator.

$P_{PTI(i)}$ is 75% of the rated power consumption of each shaft motor divided by the weighted average efficiency of the generator(s).

$P_{eff(i)}$ is the output of the innovative mechanical energy efficient technology for propulsion at 75% main engine power.

$P_{AEeff(i)}$ is the auxiliary power reduction due to innovative electrical energy efficient technology measured at *PME(i)*.

P_{AE}, is the required auxiliary engine power to supply normal maximum sea load including necessary power for propulsion machinery/systems and accommodation.

f_j is a correction factor to account for ship specific design elements.

f_w is a non-dimensional coefficient indicating the decrease of speed in representative sea conditions of wave height, wave frequency and wind speed.

$f_{eff(i)}$ is the availability factor of each innovative energy efficiency technology.

f_i is the capacity factor for any technical/regulatory limitation on capacity.

f_c is the cubic capacity correction factor and should be assumed to be one (1.0) if no necessity of the factor is granted.

f_l is the factor for general cargo ships equipped with cranes and other cargo-related gear to compensate in a loss of deadweight of the ship [23].

For each type of vessels covered by the energy efficiency index the limiting curves on carbon dioxide emissions are introduced, calculated according to the formula: $EEDI = a \cdot b^{-c}$, where *b* is the deadweight (bulk carriers, tankers, gas carriers, ro-ro cargo ships, general cargo ships) or gross tonnage (passenger ships, including the ro-ro passenger ships), *a* and *c*—coefficients the values of which are given in Table 1.7 [25].

Table 1.7 Coefficients for EEDI limit values calculation

Vessel type	Coefficient a	Coefficient c
Bulk carriers	961.79	0.477
Gas carriers	11.200	0.456
Tankers	1218.8	0.488
Containerships	174.22	0.201
General cargo ships	107.48	0.216
Refrigerator cargo carriers	227.01	0.244
Passenger ships	1218.8	0.488

The regulatory requirements to CO_2 emissions are planned to be introduced stagewise from 2013 to 2025 (Table 1.8) [25].

IMO offers to include SEEMP (Ship Energy Efficiency Management Plan) to the required documents defining the status of marine power equipment and measures for emissions reduction [26].

The possibility of introducing restrictions for other groups of emissions, primarily to PM [11, 20, 21, 27] is under discussion. If currently only the emissions of sulphur oxides from ships are under control globally, the Environmental Protection Agency in the United States (EPA—Environment Protection Agency) developed requirements for emissions from marine diesel engines including restrictions on the PM—Tier 1–4 named by analogy with the limiting curves of the Technical Code. These standards are used primarily in California with especially strict control in the coastal cities of Los Angeles and Long Beach [16].

1.3 Methods of Ship Power Plants Emissions Reduction

Methods of the emission reduction in the exhaust gases pollutants of main and auxiliary engines and boilers of the ship power plants can be divided into two categories:

- primary methods aimed to reduce the amount of harmful substances formed during the combustion process within the cylinder of the diesel engine by improving the burning process, conventional fuel upgrading and alternative fuels usage;
- secondary methods involve the removal of harmful components from the engines exhaust gases or the exhaust gases purification before releasing them into the atmosphere [28, 29].

Methods of the exhaust gases purification can be classified according to other characteristics depending on the components of the emissions, the content of which can be reduced in terms of their usage (Fig. 1.4) [1].

The basic technologies of the diesel engines nitrogen oxides and solid particles emissions reduction according to some researchers are [1, 30]:

Table 1.8 Reduction factors for the EEDI restriction curves

Vessel Type	Deadweight, t	Stage 01/01/2013–31/12/2014	Stage 11/01/2015–31/12/2019 (%)	Stage 21/01/2020–31/12/2024 (%)	Stage 3 c 1/01/2025 (%)
Bulk carriers	>20,000	0%	10	20	30
	10,000–20,000	–	0–10	0–20	0–30
Gas carriers	>10,000	0%	10	20	30
	2000–10,000	–	0–10	0–20	0–30
Tankersand combination carriers	>20,000	0%	10	20	30
	4000–20,000	–	0–10	0–20	0–30
Containerships	>15,000	0%	10	20	30
	10,000–15,000	–	0–10	0–20	0–30
General cargo ships	>15,000	0%	10	15	30
	3000–15,000	–	0–10	0–15	0–30
Refrigerator cargo carriers	>5000	0%	10	15	30
	3000–5000	–	0–10	0–15	0–30

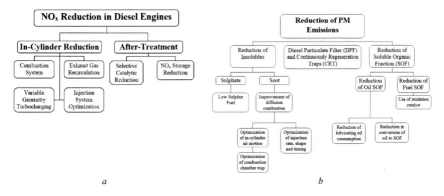

a   b

Fig. 1.4 Technologies of decreasing of the diesel engines emissions: **a** nitrogen oxides; **b** particulate matter

- high pressure of the fuel injection;
- fuel injection electronic control;
- exhaust gases recirculation;
- change of the turbochargers flow parts geometry;
- usage of filters and traps for separation of solid particles;
- catalytic reduction.

The number of solid particles and nitrogen oxides is reduced by improving the mixture formation and fuel mixture combustion process within the diesel engine. The following measures affect positively the quality of the working process [1]:

- The fuel injection high pressure together with a small-diameter hole of the injection nozzle to improve the fuel atomization. The diameter of the nozzle hole is 0.16–0.18 mm instead of the previously common 0.28–0.35 mm.
- The fuel feeding preferably to the combustion chamber part where the air is supplied, with a minimum wall wetting.
- Optimization of the supply and flow of air within the cylinder and the combustion chamber structure improvement, which provide rapid mixing of fuel and air during the entire injection process.
- Technologies of modulating control of the injection time and fuel consumption.

Let us discuss some methods of improving the environmental performance of diesel engines.

Electronic system of fuel injection. An effective achievement in the field of management of formation and access to harmful substances during combustion of fuel is the development of the engine in which the work of the fuel feeding systems and gas exchange are controlled by the processor, allowing to change the procedure of fuel supply (the law of combustion) and also gas distribution timing depending on load and speed mode. In the electronically controlled engines the operation mode with the reduced NO_x emissions is provided. The electronic management of the diesel engine systems enables more flexible management of the engine, depending on the operating mode. The electronic fuel injection system has the following properties [1]:

- the high pressure of injection (up to 200 MPa) helps the fuel to be sprayed in the form of high-dispersive droplets for quick evaporation;
- the high speed of fuel jet, which completely gets into the combustion chamber within a short period of time, efficiently using the air charge;
- precise control of the injection time;
- the fuel metering precise control, which allows monitoring the engine power and limiting the smokiness;
- the difference in the amount of fuel supplied to the various cylinders is minimized;
- the control of the initial fuel injection rate to reduce noise levels and emissions;
- sudden injection cut-off, which eliminates leaks and prevent the nozzle loading, to reduce the level of smokiness and hydrocarbon emissions.

Fuel injection slide valves. The fuel injection system optimization reduces fuel consumption and lowers emissions of particulate matter, the products of incomplete combustion. The application of the slide valves improves the fuel delivery to the cylinders and results in lower emissions of soot and nitrogen oxides. Figure 1.5 shows the fuel valves: the standard and slide types. The main difference between them is the presence of additional cavities. The slide valves differ from the standard with the injection features; they are designed to reduce leakage of fuel. The fuel injected with delay burns at a lower temperature, resulting in reduced emissions of particulate matter, volatile organic compounds and nitrogen oxides. The slide valves selection and installation requires additional investment and labor costs [31–33].

The slide valves advantages are [33]:

Fig. 1.5 Fuel injection valves: **a** standard valve; **b** slide valve

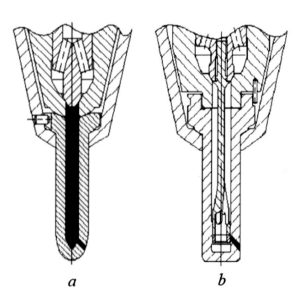

a   *b*

- fuel leakage reduction due to the smaller housing;
- reducing of soot and hydrocarbons emission;
- smokiness reduction;
- reducing of gas exhaust pipelines and surfaces and utilizing boilers gas lines loading;
- the piston head and cylinder sleeve loading reduction;
- easy installation to replace standard valves.

Water injection into the combustion chamber. The water injection into the input pipeline is an effective way to reduce emissions of nitrogen oxides. The concentration of nitrogen oxides also decreases due to water addition to the fuel. Along with a significant reduction of nitrogen oxides, the injection of water increases the content of CO and hydrocarbons in the exhaust gases, including carcinogenic benzpyrene.

The water is supplied directly to the engine cylinders. The current technology requires fresh water that demands additional water storage tanks or the capacity of desalination plants improvement. The system uses electric pumps for water injection into the combustion chamber under the pressure of 20–40 MPa (Fig. 1.6). Water is supplied before the fuel injection. The system shutdown does not lead to changes in the engine characteristics, as it is not connected with a fuel feeding system. The system installation is possible while the vessel construction. Typically, the amount of water supplied comprises 40–70% of the fuel amount, this ratio reduces NO_x formation by 50–60% [34].

Humidified air supply into the engine. The humidified air supply to the engine (Humid Air Motor—HAM) is an alternative to the water injection. The humidified air supply system uses heated boosted air saturated with water vapor, which are generally prepared by evaporation of the sea water. This technology involves the use

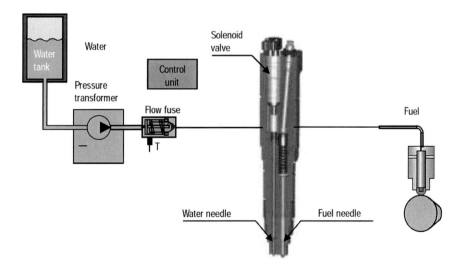

Fig. 1.6 Scheme of the water injection unit into the wärtsilä diesel engine combustion chamber

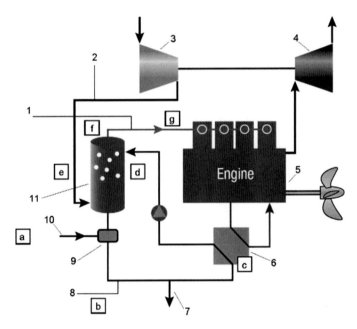

Fig. 1.7 Scheme of the HAM-technology: 1—humidified and cooled air supply; 2—heated compressed air; 3—compressor; 4—gas turbine; 5—diesel engine; 6—heat-exchange unit; 7—bleed-off; 8—water circulation; 9—water collector; 10—water supply; 11—humidifying unit

of the engine exhaust gases heat for the water evaporation. The system is combined with the engine, so the ship must have enough space to accommodate the necessary equipment. The amount of water supplied to the engine is about three times more than the amount of fuel burned. This ratio "water/fuel" leads to lower emissions of nitrogen oxides by 70–80%. To obtain the necessary amount of water vapor the existing utilizing boilers or additionally installed auxiliary boilers are used [31, 35, 36]. The scheme of the humidified air supply unit is shown in Fig. 1.7 [35].

The HAM-technology is completed in the following stages (Fig. 1.7):

1. Filtered sea water is pumped into the water collector, compensating the losses from evaporation and purging.
2. The natural circulation of water occurs within the system between the water collector and the humidifying unit.
3. The sea water is heated and then evaporates within the heat-exchange unit, which is located between the water collector and humidifying unit by means of heat from engine cooling water.
4. The steam is injected into the charged air in three stages.
5. The compressed charged air from the turbocharger bypassing the coolant is supplied over the pipelines into the humidifying unit. The charged air absorbs water within the humidifying unit. Due to the complex trajectory of water movement the impurities, including salt, fall back into the drain tank and are removed

during the formation of the brine with a certain concentration of salts, this allows to avoid salt insertion into the engine.

6. To ensure protection against insertion of tiny water droplets into the combustion chamber, moist air passes through the drip collector located in the output of the humidifying unit.
7. The humidified air saturated with steam is supplied into the engine.

The fuel injection system Common rail is a modern technology of the fuel feeding systems for diesel engines with direct fuel injection. The Common rail system pump supplies fuel at high pressure (up to 200 MPa, depending on the mode of operation of the engine) into the total fuel line (Fig. 1.8). Electronically controlled injectors with electromagnetic or piezoelectric valves inject fuel into the cylinders. Depending on the design the injectors produce from 2 to 9 injections per one cycle. The exact calculation of the injection start angle and the injection quantity allows the diesel engines to comply with the increasing environmental and economic requirements [1, 37].

Recirculation of the exhaust gases. Bypass of the exhaust gases for the absorption (recirculation) can dramatically reduce emissions of toxic substances while minimizing the deterioration of the engine effective performance and fuel economy. This method is based on the fact that there is the excess oxygen within the cylinder of the diesel engine. The replacement of the fresh air charged with the exhaust

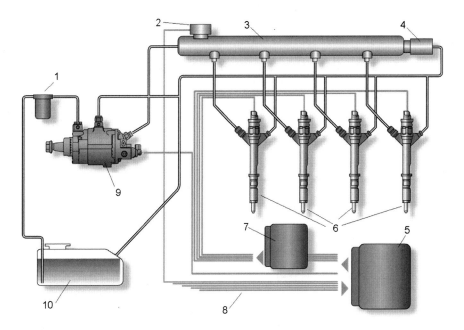

Fig. 1.8 Scheme of the fuel injection system common rail: 1 fuel filter; 2 pressure sensor; 3 high-pressure line; 4 pressure-relief device; 5 electronic control unit; 6 injector units; 7 electronic set-up unit; 8 sensors; 9 pump; 10 fuel tank

gases from the exhaust system causes the following changes in the conditions of the working process flow: reducing the amount of the working fluid within the cylinder and increase of its specific heat capacity; decrease of the oxygen concentration within the combustion chamber; compression start temperature rise; reduction of the cycle maximum temperature; reduction of the exhaust gases emissions [38].

The main advantage of this technology is decreasing the mass exchange with the atmosphere and the number of harmful emissions components changes due to recirculation, particularly in the reduction of NO_x. The disadvantages of recirculation include the difficulty of the fuel complete combustion within the cylinder in the presence of large amount of carbon dioxide and lack of oxygen. However, this disadvantage can be eliminated by the introduction of the required amount of oxygen into the working fluid [32].

Structurally, the simplest version of the exhaust gas recirculation system is a valve connecting the input and output collectors, which opens under the effect of vacuum in the input collector. For the engine stable operation in an idling mode, the system switches off.

In terms of the first method application, the exhaust gases from the turbine pressure piece are discharged and pumped into the input part of the compressor (Fig. 1.9a). Such scheme can be used in a wide range of engine loads, as it is easy to obtain the necessary pressure difference in the recirculation system valve. The disadvantage of this solution is the wear increase and reliability reduction due to the passage of exhaust gases through the compressor and the air intercooler. In addition, the bigger amount of the exhaust gases can increase the compressed air temperature to values exceeding the allowable conditions of the compressor materials strength. The exhaust

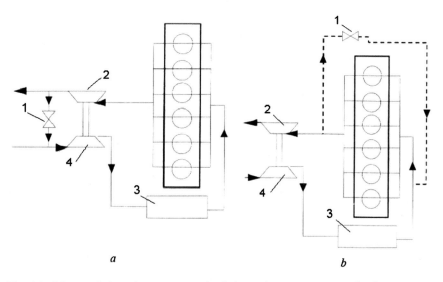

a   *b*

Fig. 1.9 Scheme of the exhaust gases recirculation unit: **a** movement under low pressure; **b** movement under high pressure; 1—recirculation system valve; 2—turbine; 3—air cooler; 4—compressor

gases impurities can form the deposits on the heat exchange surfaces of the air cooler, increasing the flow resistance and pressure loss [1].

Another method provides the exhaust gases supply from the turbine suction part into the compressor pressure part, i.e. circulating exhaust gases do not pass through the compressor and intercooler, the problems specific to the previous method, are absent (Fig. 1.9b). The oxygen content in the charged air remains constant. This method allows achieving a significant reduction in nitrogen oxide emissions without significant increase in particulate matter and smoke [1].

The additives usage. To improve the quality of conventional petroleum fuels the reactive additives and various additives (liquefied gases, alcohols and ethers, et al.) are widely used, which allows effective impact on the reactive and kinetic mechanism of the diesel engine cycle and improvement of its environmental quality.

The antismoke fuel additives are based on a variety of metals that are combustion catalysts. Adding these compounds to fuel can reduce the black smoke emissions that result from incomplete combustion. Such benefits are most significant when used with older technology engines which are significant smoke producers. There is significant concern regarding potential toxicological effects and engine component compatibility with metallic additives in general. During the 1960s, before the Clean Air Act and the formation of the U.S. EPA, certain barium organometallics were occasionally used in the U.S. as smoke suppressants. The EPA subsequently banned them because of the potential health hazard of barium in the exhaust. Smoke suppressants based on other metals, e.g., iron, cerium, or platinum, continue to see limited use in some parts of the world where the emissions reduction benefits may outweigh the potential health hazards of exposure to these materials. Use of metallic fuel additives is not currently allowed in the U.S., Japan, and certain other countries [28].

The fuel quality improvement. The fuel quality has a direct impact on the content of the exhaust gases non-standardized harmful components, and depending on the composition can influence the emission of standardized components. There is a direct dependence between the sulphur in the fuel and the sulphur oxides and particulate matter in the exhaust gases. The content of cyclic and polycyclic aromatic hydrocarbons in the fuel increases the smoking of the exhaust gases and so on. The effect on the NO_x emissions is directly shown through the organic nitrogen compounds contained in fuel. The indirect influence on the formation of fuel NO_x through the fuel burning rate and the flame temperature is also possible. However, the fuel NO_x is specific only for heavy fuels and make up no more than 5% of the thermal oxide. The studies show that the cumulative effect of reduction of harmful emissions produced by improving the quality of traditional fuels is sufficient enough and is within 10–20% [39].

Alternative fuels. The alternative fuels usage can dramatically affect the emission performance [40–44]. This is one of the possible directions of engines modernization, promising a significant effect, particularly in reduction of the NO_x emissions [45]. In the process of assessing of the prospects for the usage of alternative fuels it is necessary to consider the significant material costs for the engines modernization, the organization of their production and the infrastructure creation for the operation and service. Environmental effect from the use of the alternative fuels will be discussed in detail below.

Secondary methods of engine exhaust gases purification can be divided into three main groups:

- absorption (absorption of the individual components of the exhaust gases with the volume of liquid);
- adsorption (absorption of the individual components of the exhaust gases with the solids);
- catalyst (conversion into other chemical compounds).

The company Advanced Cleanup Technologies Incorporated (ACTI), California, U.S., specializes in the management of hazardous emissions and developing methods to reduce emissions from ships docked in port. It uses an approach called "Actual Marine Emissions Control System" (AMECS) providing the use of purification technologies in the scrubbers and catalytic reduction. The AMECS system can be located on a barge or directly in the port (Fig. 1.10). The uniqueness of this approach consists of the fact that the system uses adjustable receiving umbrellas, which are installed on the engine gas outlet. While the ship is in the port, the exhaust gases are discharged via a pipeline linking to umbrellas and off-gas treatment system from harmful substances [31].

Currently, there are only experimental devices, but the practical application of the AMECS system showed the effectiveness of the emissions reduction technology. It allows ship owners to reduce emissions without modernization of vessels and can be used for any type of vessels. The disadvantage is that the emissions reduction does not occur in the areas of transit passage of ships and other ports that are not equipped with such systems [31].

Scrubbers. The exhaust gases treatment technology within the scrubbers is based on the ability of water with a high pH level (with a high content of alkalis) neutralize the acid and thereby reduce emissions of sulphur. The sulphur oxides react with water

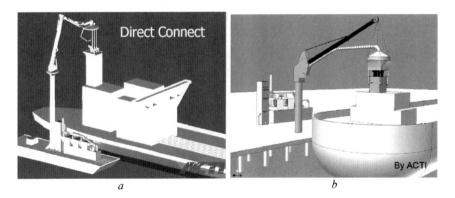

a   *b*

Fig. 1.10 Ship Exhaust Gases Purification at a Port: **a** AMECS system on the barge; **b** AMECS system on the port territory

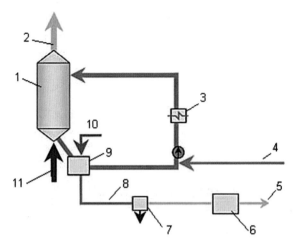

Fig. 1.11 Scheme of the exhaust gases purification within the scrubbers: 1—scrubber; 2—exhauster; 3—cooler with sea water; 4—alkali injection (caustic soda); 5—water discharge (only in the open sea); 6—collecting tank; 7—sludge tank; 8—contaminated sea water treatment; 9—treatment tank; 10—fresh water; 11—engine or boiler exhaust gases

to form acids, which are neutralized with alkali. The sea water with a high alkalinity or fresh water with an alkali solution can be supplied into the scrubber. Figure 1.11 shows the exhaust gas purification circuit from sulphur within the scrubbers with fresh water by Wärtsilä [31].

After the water mixing with the exhaust gases the particulate matter of the sulphur dioxide is carried away, and the pre-purified water can be discharged overboard, because oceans and seas are natural sulphur storages and contain a significant amount of it. The sea water usage greatly simplifies the process flow. There is no need in additional amounts of fresh water and chemical additives. The drawback is the need to supply significant amounts of water with the required alkalinity level, which may vary depending on the navigation area. The advantage of the fresh water usage is the possibility of stable water supply with the required predetermined pH level and deep removal of sulphur, which is equivalent to using fuel with $S = 0.1\%$ [31].

Desulphurization in scrubbers is an alternative to using low sulphur fuel. Application of the system allows the vessel to burn fuel with a high sulphur content, which lowers maintenance costs. The technology of the exhaust gases "washing" reduces SO_2 emissions by 69–94% on vessels and reduces the amount of particulate matter.

The main expenses of the technology are the initial capital investments. Operational and maintenance costs include the removal of the sludge from the discharged overboard sea water and maintenance of the pumps. In terms of the small difference between the prices of the fuel a high sulphur content and low the costs can increase and even make its use uneconomical [31, 46].

Catalytic reduction. The process of selective catalytic reduction comprises of the reagent injection (ammonia or urea) in the exhaust gases flow, as this takes place the

content of nitrogen oxides of the gases going through the catalyst is reduced by 90%
(Fig. 1.12). This unit has a considerable weight and overall dimensions and therefore
it would be preferable to install it on the newly constructed ships, than upgrade
existing ones. Most catalytic reduction systems are used on four-stroke engines [31].
The engine exhaust gases temperature is to be above 270 °C for efficient operation
of the technology.

The catalyst must be changed periodically, as it is sensitive to the pollutants of the
exhaust gases and disintegrates over time. Soot, alkali, phosphorus oxides, sulphur

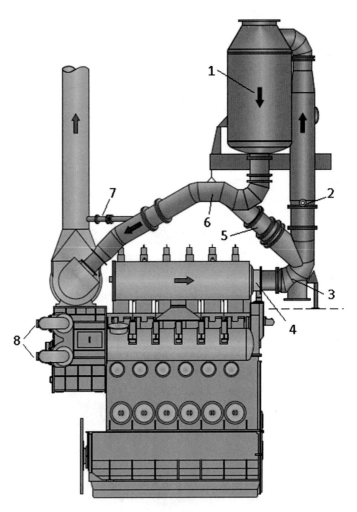

Fig. 1.12 Scheme of the catalytic reduction system: 1—reactor for catalytic reduction; 2—injector
for urea supply; 3—temperature sensor upstream the reactor; 4—additional flange in the exhaust
collector; 5—bypass pipeline (bypass); 6—temperature sensor downstream of the reactor; 7—
bypass device of the turbocharger; 8—blowers additional drives

compounds, that normally present in the engine exhaust gases can deactivate the catalyst, which durability can be extended by using the fuel with lower sulphur content. The usage of urea in the system leads to decomposition of NO_x to nitrogen and water. Reducing the NO_x content up to 90% is achieved by ammonia supply at 15 g/(kW·h). The excess catalyst results in a so-called "slip ammoniac", when ammonia, not reacted with the nitrogen oxides, goes with the exhaust gases of the engine. Note that ammonia is an air pollutant and can cause corrosion of the output system elements [1, 31, 47].

Shore power supply. An effective way to reduce emissions is to stop the marine engines operation when entering and parking at the port and the ship exhaust gases purification when docked in the port. Upon the engines operation termination the ship is provided with the energy from the shore, the level of pollution reduction depends on the type of vessel and the duration of the engines operation in the port. The emissions reduction is defined as the pollution of the port area, which takes place while the engines operation, dispose of the emissions, which is formed when the energy supply of the ship is executed from the shore. The cost of this service depends on the type of vessel, the capital expenses account for 1.2–6.7 million dollars, they involve additional investment in the construction of the ship, a special terminal for power supply and serving barge (Fig. 1.13) [48].

The fuel consumption reduction depends on the engine shutdown time. The potential difficulty is the ability to satisfy the significant power needs of ships by means of the shore-based facilities, as this requires creation of appropriate infrastructure in the ports and modernization of vessels for the installation of the necessary equipment in order to ensure proper maintenance and security in terms of the high voltage cables usage. This technology has been applied since the late 1980s in a number of countries. For its wide distribution, there is a need to standardize the system requirements

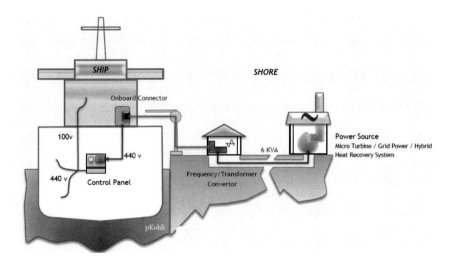

Fig. 1.13 Scheme of the shore ship power supply

Table 1.9 The emissions and fuel consumption level change under different emissions reduction technologies application

Emissions reduction methods	NO_x	PM	Fuel consumption	CO_2	SO_x
Engines exhaust gases treatment					
Catalytic converter (%)	−81	−35	−7.5	−7.5	−7.5
Soot collection filters	–	−85%	+2%	+2%	+2%
Engine operation process control					
Common rail system engines (%)	−10	−10	−10	−10	−10
Fuel change					
Biodiesel fuel B100 (%)	+ 10	−30	+15	−65	−100
B20 (80% of diesel fuel and 20% of biodiesel fuel) (%)	+ 2	−6	+3	−13	−20
Diesel fuel with sulphur content up to 0.5%	–	−17%	–	–	−100%
New types of engines					
Engines operating on natural gas (%)	−98.5	−97.5	+ 4.5	−10	−100

to ensure the reliable and easy connection of the ship power system with the local power systems [31, 48].

Such a variety of emissions reduction methods leads to the complexity of a particular technology choice, which is dependent on many factors that determine the effectiveness of the method in the specific conditions. The effectiveness of measures to improve the environmental engine performance is usually evaluated by two parameters: the level of the emissions reduction and the various groups of expenses (capital investments and operating costs). Table 1.9 shows a comparison of different groups of emissions reduction technologies for the ships of the interior sailing comparing to the ship power plants operation without any emissions reduction technologies application [49]. According to these data the most effective method of the exhaust gases basic harmful components reducing is the use of gas reciprocating engines, fueled by natural gas.

Tables 1.10, 1.11, 1.12 provide a comparison of various methods to reduce emissions of internal combustion engines for marine eco-efficiency, the application field and status of the technology is given, as well as the main producers. Table 1.10 information is represented according to Entec (2005), Eyring et al. (2005), in Table 1.11—Entec (2005), Eyring et al. (2005), Genesis Engineering (2003), Wärtsilä's (2006), in Table 1.12—Entec (2005), Eyring et al. (2005), CARB (2005) data [5]. Table 1.13 along with the environmental effectiveness of various methods of emissions reducing represents information on the payback period of the technologies [31].

Comparison of the data in Tables 1.9, 1.10, 1.11, 1.12 and 1.13 indicates that the efficiency of emissions reduction methods vary depending on the company that

Table 1.10   Sulphur oxides reduction technologies comparison

Technology	SO_x reduction, %	Field of application		Technology status	Producers
		Engines	Vessels		
Heavy fuel with S = 1.5%	44	2- and 4-stroke engines	All types	Commercially-available	Oil refining enterprises
Heavy fuel with S = 0.5%	81	2- and 4-stroke engines	All types	Commercially-available	Oil refining enterprises
Scrubbers application	75	2- and 4-stroke engines	All types	Demonstrational or customized	Marine exhaust solutions BP-Kittiwake

Table 1.11   Nitrogen oxides reduction technologies comparison

Technology	NO_x Reduction, %	Field of application		Technology status	Producers
		Engines	Vessels		
Engine modernization	20–30	2- and 4-stroke engines	All types	Some models are standardized, the rest are expected to be produced in 5–10 years	Caterpillar/MaK, MAN B&W, Wärtsilä
Catalytic converters	8–95	4-stroke medium- and high-speed engines, some 2-stroke engines of new generation	All types	Commercially available	ArgillonGmbh, Munters, Wärtsilä
Water-fuel emulsions	0–30	2- and 4-stroke engines	All types	Demonstrational or customized	MAN B&W, MTU, Orimulsion, PuriNO$_x$
Water injection into the combustion chamber	50	4-stroke medium-speed engines	With the Wärtsilä engines	Commercially available	Wärtsilä
Humidified air supply into the engine	70	4-stroke engines	Has been demonstrated on ferries	Only demonstrational sample	Munters, MAN B &W
Exhaust gases recirculation	35	4-stroke engines	–	Research and experimental implementation	MAN B&W

Table 1.12 Emissions reduction integrated technologies comparison

Technology	Reduction, %	Field of application		Technology status	Producers
		Engines	Engines		
Marine diesel fuel with S = 0.5%	SO_x: 75, PM: 80	2- and 4-stroke engines	All types	Commercially available	Oil refining enterprises
Marine diesel fuel with S = 0.1%	SO_x: > 90, PM: > 80	2- and 4-stroke engines	All types	–	Oil refining enterprises
Marine gas oil	SO_x: > 90, PM: > 80	2- and 4-stroke engines	All types	Commercially available	Oil refining enterprises
Shore ship power supply	PM, NO_x, SO_x: > 90	2- and 4-stroke engines	All types, currently are used on ferries, cruise ships, tankers and Ro-Ro ships	Commercially available, currently is installed in Sweden, the USA (Los-Angeles, Juneau, Seattle)	Cavotech
Vessels exhaust gases purification system at a port	PM, NO_x, SO_x: > 90	2- and 4-stroke engines	All types	The first demonstration at the port of Long-Beach	ACTI

conducted the study. Cost-effectiveness of emissions reduction technologies is given in Tables 1.14, 1.15 and 1.16. Table 1.14 shows the expenses on the engines of small, medium and high power [5, 31].

The smallest specific capital expenses occur while the slide valve application for the diesel engine fuel injection system that results in reduced emissions of nitrogen oxides, soot and volatile compounds (Table 1.13). The largest specific capital expenses occur while humidified air supply and secondary methods of the exhaust gases purification application. These technologies, without being packaged ones, provide a significant reduction of the nitrogen and sulphur oxides emissions. Moreover the operating expenses for the implementation of the humidified air supply technology is much smaller comparing to the same values for the other methods.

According to Table 1.15 data the highest capital expenses are spent on scrubbers usage and humidified air supply to the engine. The technologies on the exhaust gas catalytic reduction systems development and vessels power services from the shore at ports systems development are quite expensive (Table 1.16). The most expensive in operation is the method of ships power supply from the shore, and the estimated values of the different researchers can vary by almost three times.

Table 1.13 Emissions reduction with the engines exhaust gases while different methods application

Emission control technologies	Emission reductions, %					Change in fuel consumption, %	Lifespan, years
	NO$_x$	PM	SO$_2$	CO	VOC*		
Fuel injection: slide valves	20	25–50	0	0	<50	0	2.5
Advanced operational modifications	30	0	0	0	0	0	25
Exhaust gas recirculation	35	0	0	–	–	0	
Humid air motor	70	0	0	0	0	0	12–15
Water emulsion (30% water)	30	0	0	0	0	>3	–
Water injection	50	0	0	0	0	0	25
Selective catalytic reduction	90	0	0	0	0	0	15
Seawater scrubbing	0	25	69–94	0	0	0	15
Shore power (advanced maritime emissions control system)	95	99	99	–	50	0	10
Switching fuels from 2.7% to 1.5% Sulphur content	0	18	44	0	0	–	–
Switching fuels from 2.7% to 0.5% Sulphur content	0	20	81	0	0	–	–

1.4 Conclusions

1. Water transport contributes significantly to air pollution, especially nitrogen oxides and sulphur emissions which for 2010 are about 27% and 10% of the global total emissions, respectively. In the future according to the researchers the

Table 1.14 The costs of different methods of the exhaust gases emissions reduction

Method	Method expenses, USD / Method costs, USD/kW								
	Capital (new construction)			Annual operation and maintenance expenses			Capital (modernization)		
	Engines power, kW								
	3000	10,000	25,000	3000	10,000	25,000	3000	10,000	25,000
Slide valve usage in the fuel injection system	10,920 / 3.64	36,400 / 3.64	91,000 / 3.64	–	–	–	–	–	–
Engine structure change	134,105 / 44.70	149,705 / 14.97	215,725 / 8.63	–	–	–	–	–	–
Humidified air supply	578,500 / 192.83	1,615,500 / 161.55	3,430,000 / 137.20	2950 / 0.98	9575 / 0.96	23,900 / 0.96	578,500 / 192.83	1,740,500 / 174.05	4,055,000 / 162.20
Water injection into the combustion chamber	169,665 / 56.56	338,223 / 33.82	686,166 / 27.45	41,488 / 13.83	135,700 / 13.57	338,750 / 13.55	–	–	–
Catalytic reduction	282,438 / 94.15	656,763 / 65.68	1,509,254 / 60.37	169,400 / 56.47	427,576 / 42.76	1,001,500 / 40.06	423,656 / 141.22	985,144 / 98.51	2,263,880 / 90.56
Scrubbers usage	523,320 / 174.44	1,687,560 / 168.76	4,233,600 / 169.34	15,700 / 5.23	33,751 / 3.38	42,336 / 1.69	747,600 / 249.20	2,410,800 / 241.08	6,048,000 / 241.92
Fuel sulphur content reduction from 2.7% to 1.5%	0	0	0	196,134 / 65.38	642,118 / 64.21	1,602,796 / 64.11	–	–	–
Fuel sulphur content reduction from 2.7% to 0.5%	0	0	0	252,171 / 84.06	825,580 / 82.56	2,060,738 / 82.43	–	–	–

Table 1.15 Ships emissions reduction technologies capital expenses

Technology	Data source and technology costs				
	ENTEC (2005), EUR/kW	ENTEC (2005), USD/kW	US EPA (2003), USD/kW (medium-term forecast)	US EPA (2003), USD/kW (medium-term forecast)	GENESIS ENGINEERING (2003), USD/kW
NO_x.reduction					
New engines with the slide valve (baseline models)	0.29	0.4	–	–	–
Old engines with the slide valve (baseline models)	0.42	0.5	–	–	–
Engines advanced models	6	7	8	3	
Water injection into the combustion chamber	19	24	20	13	12
Humidified air supply	113	141			
Catalytic reduction	63	78	54	50	49
SO_xreduction					
Scrubbers use	168	209	–	–	–
Integrated Emissions Control ($SO_x + NO_x + PM$)					
Shore ship power supply (new construction)	55	68	–	–	–
Shore ship power supply (existing systems modernization)	78	97	–	–	34

emissions level of ship power plants will grow. According to forecasts, by 2020, emissions of nitrogen and sulphur oxides from shipping in EU countries will have exceeded the similar value of the land-based sources, including stationary power plants, and land transport.

2. The high part of emissions is accounted for the transport ships, equipped mostly with diesel engines, the use of which leads to significant, compared to the steam

Table 1.16 Ship emissions reduction technologies operation expenses

Technology	Data source and technology costs				
	ENTEC (2005), EUR/(MW · h)	ENTEC (2005), USD/(MW · h)	US EPA (2003), USD/(MW · h) (medium-term forecast)	US EPA (2003), USD/(MW · h) (long-term forecast)	GENESIS ENGINEERING (2003), USD/(MW · h)
NOₓreduction					
New engines with the slide valve	0	0	–	–	–
Old engines with the slide valve	0	0	–	–	–
Engines advanced models	0	0	0	0	–
Water injection into the combustion chamber	2	3	1	1	3
Humidified air supply	0.15	0.19	–	–	–
Catalytic reduction (heavy fuel S = 2.7%)	6.2	8	–	–	–
Catalytic reduction (heavy fuel S = 1.5%)	4.5	6	9.5	9.4	–
Catalytic reduction (light fuel use)	3.4	4	–	–	19
SOₓreduction					
Scrubbers use	0.3	0.4	–	–	–
Fuel sulphur content reduction from 2.7% to 1.5%	10	12	–	–	4
Fuel sulphur content reduction from 2.7% to 0.5%	13	16	–	–	20

(continued)

Table 1.16 (continued)

Technology	Data source and technology costs				
	ENTEC (2005), EUR/(MW · h)	ENTEC (2005), USD/(MW · h)	US EPA (2003), USD/(MW · h) (medium-term forecast)	US EPA (2003), USD/(MW · h) (long-term forecast)	GENESIS ENGINEERING (2003), USD/(MW · h)
Integrated Emissions Control (SO$_x$+ NO$_x$+ PM)					
Shore ship power supply	16.3	20	–	–	57.1

turbine units and gas turbine engines, emissions of nitrogen oxides, carbon monoxide, and volatile organic compounds. Along with the standardized levels of nitrogen and sulphur emissions from the ship power plants the dangerous components for human health and environmental get into the atmosphere, including and greenhouse gases.

3. The emissions of nitrogen and sulphur oxides from ships are standardized at the international and regional levels, since 2012 it is planned to introduce a limit on emissions of carbon dioxide, the ability to limit other harmful components at the global level is under consideration.

4. The implementation of the restrictions on the sulphur content in marine fuel (not more than 0.5%) at the global level will make it possible to reduce sulphur emissions by 78% by 2020, similar activities only in the areas of SECAs will make it possible to reduce total SO$_2$ emissions by almost 15%.

5. The most environmentally efficient method of the ships emissions reduction is the use of the exhaust gases purification systems, when the ships are docked at the port, providing almost complete removal of the main harmful components. According to various researchers, this method is quite expensive in construction and the most expensive while operation compared to the basic primary and secondary technologies of the harmful emissions reduction.

6. One of the economically feasible ways to improve the eco-efficiency and energy security of ships is the adoption of alternative fuels that reduce harmful emissions of marine engines into the environment.

References

1. B. P. Pundir, Engine emissions: pollutant formation and advances in control technology. Oxford: Alpha Science International Ltd, 2007, 301 p.
2. A. S. Hachiyan, K. A. Morozov, V. N. Lukanin, Dvigateli vnutrennego sgoraniya [Internal Combustion Engines]. Moskva: Vysshaya shkola Publ., 1985, 311 p.
3. Ya. B. Chertkov, Motornye topliva [Motor Fuels]. Novosibirsk: Nauka Publ., 1987, 207 p.

4. US Environment protection agency. Greenhouse gas properties [Online]. Available: https://www.epa.gov/outreach/scientific.html.
5. A. Friedrich, F. Heinen, F. Kamakate, D. Kodjak (2007). Air pollution and greenhouse gas emissions from ocean-going ships: impacts, mitigation options and opportunities for managing growth. Presented at the international council on clean transportation [Online]. Available: https://www.georgiastrait.org/files/share/PDF/MarineReport_Final_Web.pdf.
6. N. V. Lazareva, E. N. Levinoy, Vrednye veshchestva v promyshlenosti. Spravochnik dlya himikov [Industry Harmful Substances]. Lvov: Himiya Publ., 1976, p. 592.
7. C. Trozzi, R. Vaccaro (1998) Methodologies for estimating future air pollutant emissions from ships [Online]. Available: https://www.researchgate.net/publication/259470337_Methodologies_for_estimating_air_pollutant_emissions_from_ships_a_2006_update.
8. The Swedish NGO secretariat on acid rain. Cost-benefit analysis of using 0, 5% marine heavy fuel oil in European sea areas [Online]. Available: www.airclim.org/reports/cba_briefing_jan05.pdf.
9. A. McKinnon, M. Piecyk (2011, Jan) Measuring and managing CO2 emissions of European chemical transport [Online]. Available: https://www.cefic.org/Documents/IndustrySupport/Transport-and-Logistics/MeasuringAndManagingCO2EmissionOfEuropeanTransport-McKinnon-Report%20-24.01.2011.pdf.
10. H. N. Psarafitis, C. A. Kontovas, "CO_2 emission statistics for the world commercial fleet," WMU Journal of Maritime Affairs, pp. 1–19, 2009.
11. "Controlling emissions in two-stroke marine diesel," MER, pp. 16–21, Nov. 2008.
12. "New legal driver for emission monitoring," The Naval Architect, p. 18, Sep. 2007.
13. S. Ortmanns (2007) Economic instruments for reducing air pollution from ships. Presented at CCB Annual Conference [Online]. Available: https://www.ccb.se/documents/airpollredshipsSO.pdf.
14. "Ships' emissions reach crucial stage," The Naval Architect, pp. 39–45, March. 2008.
15. V. V. Voznitskiy, Praktika ispolzovaniya morskih topliv na sudah [Ships Marine Fuels Use]. Saint Petersburg: Biblioteka sudovogo mehanika Publ., 2006, 124 p.
16. Sacramento: California Environment protection agency (2010) Initial statement of reasons for the proposed rulemaking: staff report [Online]. Available: https://www.arb.ca.gov/regact/2010/chc10/harborcraftisor.pdf.
17. "GL cleans the air," The Naval Architect, p. 75, Sep. 2008.
18. "Mandatory CO2 Index takes step closer," Shipping World & Shipbuilder, p. 8, July/Aug. 2008.
19. "Oslo meeting prepared GHG reduction mechanism," MER, pp. 46–47, Sep. 2008.
20. L. Schumacher, A. Chellappa, W. Wetherell, M. D. Russell, "Scientific IMO SOx study completed," MER, p. 20, Feb. 2008.
21. "Shipping emissions remain burning issue," The Naval Architect, p. 77, Jan. 2008.
22. "The potential to cut CO_2," The Naval Architect, pp. 30–34, June 2009.
23. IMO, MEPC.245(66), Annex 5 (2014) Guidelines on the Method of Calculation of the Attained Energy Efficiency Design Index (EEDI) for New Ships [Online]. Available: https://www.imo.org/en/OurWork/Environment/PollutionPrevention/AirPollution/Documents/245(66).pdf.
24. "The ultra-long-stroke engine and Handymax tanker propulsion," DIESELFACTS, no 3, pp. 6–8, 2011.
25. Indian Register of Shipping (2015) Implementing Energy Efficiency Design Index. [Online]. Available: https://www.irclass.org/files/marine_publications/EEDI_2015.pdf.
26. M. Simms, "GHG emissions and legislation," MER, pp. 42–44, May 2011.
27. "Politicians to act on emissions," MER, pp. 40–42, Apr. 2008.
28. Chevron. Diesel Fuels Technical Review [Online]. Available: https://www.chevronwithtechron.ca/products/documents/Diesel_Fuel_Tech_Review.pdf.
29. "Emission control, two-strokes," Propulsion, pp. 18–22, 2007.
30. "CIMAC Circle talks emissions," MER, p. 34, Nov. 2008.
31. A. Fournier (2006, Feb) Controlling air emission from marine vessels: problems and opportunities [Online]. Available: https://fiesta.bren.ucsb.edu/~kolstad/temporary/Marine_Emissions_2-11-06_.pdf.

32. O. Masaki (2007) Possible change of ship design including engine room about emissions of NOx and SOx [Online]. Available: https://www.nmri.go.jp/main/cooperation/imo_iso/con tents/IMO2007/airpolws/presentation/airpolws4.pdf.
33. MAN B&W Diesel R&D (2006) Retrofit control technology discussion [Online]. Available: https://www.fasterfreightcleanerair.com/pdfs/Presentations/FFCACA2006/Kjeld%20A abo%20-%20Low%20Emission%20Technology%20Modifications%20to%20Ships.pdf.
34. Wärtsilä (2011) Technology review of Wärtsilä 46 engine [Online]. Available: https://www.die selduck.ca/machine/01%20prime%20movers/rhapsody%20de1/Wartsila%20W46.pdf.
35. MAN Diesel SE. Humid air motor: technology for green profits [Online]. Available: https:// mandieselturbo.com/files/news/filesof15316/3-12%20HumidAirMotor.pdf.
36. D. Tinsley, "Clean, green and lean," Shipping World & Shipbuilder, pp. 20–27, Nov. 2008.
37. K. Karsdash, "Retrofit revolution," Shipping World & Shipbuilder, p. 31, Sep. 2011.
38. "Copenhagen test centre brings NOx to its Knees," DIESELFACTS, no 3, pp. 11–12, 2007.
39. O. A. Gladkov, Sozdaniye malotoksichnyh dizley rechnyh sudov [Low-toxic Diesel Fuels Development for River Ships]. Lvov: Sudostroyeniye Publ., 1990, 112 p.
40. "Four-stroke technology and a better environment," DIESELFACTS, no 3, p. 8, 2007.
41. K. Karsdash, "DK Group welcomes energy index," MER, pp. 14–15, June 2011.
42. "Keeping up to speed with LNG technology," Offshore Marine Technology, no 2, p. 14, 2011.
43. "Maintaining a competitive edge," Shipping World & Shipbuilder, pp. 30–33, Dec/Jan. 2011.
44. "The challenge of cutting CO_2," Propulsion, pp. 4–12, 2011.
45. "Cleaner four strokes," Propulsion, pp. 42–44, 2007.
46. P. Baan, "Controlling emissions," Propulsion, pp. 14–18, 2011.
47. "The unseen emission—revealed as a threat," MER, pp. 40–42, Sep. 2008.
48. P. Kohli. Fuel—energy in ports maritime industry. Cold ironing an overview profits [Online]. Available: https://crosstree.info/Documents/ColdIroning.pdf.
49. "Improving performance on rivers and canals," Propulsion, pp. 44–48, 2008.

Chapter 2
Modern State of Using Alternative Fuels in Marine Engineering

2.1 Impact of Fuel Characteristics on the Parameters of Fuel Systems and Ship Power Plants

Increase in the SPP fuel systems efficiency depends on many factors:

- the use of high-quality fuels with basic performance that meets standards;
- the system supply with the same type of equipment with optimal characteristics based on its unification and typing;
- availability of the deep-laid fuel purification system and fuel preparation technology; the systems energy consumption reduction by saving energy and fuel;
- if possible, minimization of the mass-size and cost parameters of the equipment;
- increase in the systems reliability;
- reduction of the environmental pollution or its prevention;
- automation and timely diagnostics of the systems [1, 2].

The efficiency of the fuel systems operation is largely determined by the quality of fuel taken on board of the vessel [3]. The order of fuels should be carried out taking into account the recommendations of the engine-building company, and also taking into account the possibility of the ship's fuel preparation system. The fuel selected must meet international or regional quality standards.

It is necessary to ensure mandatory sampling of fuels for the analysis of their characteristics in specialized laboratories.

The most famous international standards for petroleum fuels are: *International Standards Organization (ISO) Petroleum products—Fuels (class F); Specifications of marine fuels—ISO 8217:2005; British Standards Institution Petroleum fuels for marine oil engines and boilers BSMA 100:1989*.

The European Standards Organization has developed the standard *EN14214* applied for the biodiesel fuels, also there are other standards, such as *EN590* (or *EN590:2000*) and *DIN51606*; in the USA the standard *ASTM D6751* is applied for biodiesel fuels.

© Shanghai Scientific and Technical Publishers 2021
X. Yang et al., *Alternative Fuels in Ship Power Plants*,
https://doi.org/10.1007/978-981-33-4850-9_2

A large number of standards has been developed for the liquefied natural gas as well. They are *ASTM*, *NFPA*, *SAE*, *FTA*, *GPA* (USA), *EN* (EU), *BS* (Great Britain), *CSA* (Canada), *DNV* (Norway), *JGA* (Japan), *PB* and *RD* (Russia). The current USA standards *NFPA 59A29*, *NFPA 57*, European standards *EN 1473*, *EN 1160*, *EEMUA 14731*and other can be given as the example.

Despite all the positive qualities of alternative fuels, they should be used only when it is economically feasible. In this regard, it is necessary to identify clearly the segments of the fleet, where the use of alternative fuels is more preferable than the operation of the SPP working on petroleum fuels, heavy or light.

The analysis of literature data and publications made it possible to assess the influence of physicochemical parameters of fuels on the ship power plants parameters (Tables 2.1 and 2.2) [1, 4, 5].

2.2   Alternative Fuels Technologies

There are three main groups of alternative motor fuels [6]:

- synthetic liquid fuels obtained from non-traditional organic raw materials and similar in performance to petroleum fuels;
- mixtures of petroleum fuels with oxygen-containing compounds, such as alcohols, ethers, water, which are close to traditional petroleum fuels in performance properties;
- fuels of non-oil origin, differing in their properties from traditional (alcohols, compressed natural gas, liquefied gases).

The use of the latter group of fuels requires the modification of engines and fuel storage systems.

The main properties of the alternative fuels are described below. According to the properties, there are examples of usage in power production and transport industries, apart from the fuels that are promising for use in SPP which are considered in the Sect. 2.3.

Alcohol fuels. The main advantage of alcohols is high detonation resistance, which determines their preferential use in spark-ignition internal combustion engines as an alternative to gasoline. The efficiency coefficient (EC) of the engine operation on alcohols is higher than on oil fuel. This is due to the lower temperature of the exhaust gases, intensive removal of heat from the cylinders and more complete combustion. When using alcohol fuels, the content of the main toxic components in the exhaust gases of the engine is reduced. Because of the lower combustion temperatures of alcohols, as compared to gasoline, significantly less nitrogen oxides are released. Due to the increase in the completeness of combustion of alcohols due to the oxygen included in their composition, emissions of CO and carcinogenic aromatic hydrocarbons are reduced.

Along with the generally positive ecological efficiency, the use of alcohol fuels is accompanied by an increase in the concentration of aldehydes in the exhaust gases.

Table 2.1 Influence of the fuels' parameters on the SPP parameters

Fuels parameters	SPP Parameters of								
	Energy efficiency	Mass	Dimensions	Reliability	Durability	Self-sufficiency	Flexibility	Economical efficiency	Environmental friendliness
Density, kg/m³	+	*	*	+		+	+	+	+
Kinematic viscosity, mm²/s	+			+			+		+
Flash point, °C					*				
Cloud point, °C	+			+			+		
Boiling point, °C					+				
Filter clogging point, °C	+			+			+		
Sulphur content, %				+				+	*
Ash content, %				+					*
Vanadium and sodium content, %				+					+
Carbon content, %				+			+		+
Aluminosilicates content, %				+					
Mechanical impurities content, %				+					
Water content, %	+	+	+	+					+
Asphaltenes content, %				+					+

(continued)

Table 2.1 (continued)

Fuels parameters	SPP Parameters of								
	Energy efficiency	Mass	Dimensions	Reliability	Durability	Self-sufficiency	Flexibility	Economical efficiency	Environmental friendliness
Lower heating value, MJ/kg	*	*	*			+	+	*	
Stability	+			+			+		
Cetane number	+			+			+		
Acid number	+			+			+		
Copper-plate test				+					
Lubricity				+				+	
Propensity to microbiological contamination				+					
Stoichiometric amount of air required for complete fuel combustion, kg/kg									+
Peak flame temperature, °C				+					+
Environmental impact								*	*

(continued)

Table 2.1 (continued)

Fuels parameters	SPP Parameters of								
	Energy efficiency	Mass	Dimensions	Reliability	Durability	Self-sufficiency	Flexibility	Economical efficiency	Environmental friendliness
Storage and preparation conditions			+	+				*	
Industrial production output						+		+	
Price								*	
Human health safety records					*				

+the parameter indirectly affects the SPP parameter
*the parameter directly affects the SPP parameter (can be determined by calculation)

Table 2.2 The nature of interrelation of the SPP and fuels' parameters

SPP parameters	Degree of impact of the fuels' parameters
Energy efficiency	*Density.* The change in density can affect the value of the specific effective fuel consumption (the mass cyclic fuel supply changes) *Viscosity.* Change in viscosity can lead to difficulty in the process of fuel feed *Pour point* affects the density and viscosity, which depend on temperature; at a high pour point, the fuel supply may be disrupted unless a heating system is provided *Filter clogging point.* At low temperatures, the "plugs" are formed in the fuel lines, and filters are clogged, which prevents the fuel from advancing, resulting in engine stops *Water content* reduces the heat of combustion of the fuel volume unit, thereby reducing the efficiency factor of the SEP *Lower heating value.* The lower this value is, the greater the fuel supply is to ensure the nominal power of the engine, which is especially important while switching to another type of fuel *Stability.* The unstable fuel density, viscosity, composition and other properties can vary *Cetane number.* Awareness of the CN is necessary to determine the feasibility of using the fuel under consideration for the specific engine in order to obtain preliminary information on how the fuel combustion will take place *Acid number* can affect the cetane number
Mass and dimensions	*Density.* It affects the volume of fuel reserves. It is an important indicator when considering the limited space onboard of the ship
	Water content of the fuel reduces the useful volume of the tanks and increases the mass of fuel stocks
	*Lower heating value.*The lower the combustion temperature is, the greater the required amount of fuel on the vessel is
	Storage and preparation conditions. They affect the composition, overall dimensions and mass of fuel system elements
Durability	*Flash point.* The higher the flash point is, the less is the risk of fire and explosion of fuel in emergency situations
	Boiling point. There is a danger of fuel vapor accumulation in the engine room at a low boiling point. As a result, the explosion or people poisoning can happen
	Human health safety records. They determine the fuel hazard to humans in terms of inhaling the fuel vapors (toxicity) and contact with open skin areas
Self-sufficiency	*Density.* It affects the volume and weight of fuel reserves, the increase of which can affect the duration of the operation modes
	*Lower heating value.*The lower the combustion temperature is, the greater is the required fuel stock on the ship. This can affect the duration of operation and parking modes
	Industrial production output. It is necessary to take into account the possibility of fuel bunkering in ports within the vessel navigation area

(continued)

Table 2.2 (continued)

SPP parameters	Degree of impact of the fuels' parameters
Cost	*Density.* When using high-density fuels, additional equipment (heaters, additional filters, etc.) is needed. This increases the construction cost of the vessel
	Sulphur content. Burning of fuels with high sulfur content leads to intensive wear of engine parts, so the cost of the parts repair and replacement increases significantly. In terms of long-term storage, some sulfur compounds accelerate the formation of tar, and it reduces the fuels' combustion temperature
	Lower heating value. The lower the combustion temperature is, the greater the mass of fuel reserves is, and consequently, the higher the costs are
	Lubricity. The low lubricating properties of fuel lead to early wear of the equipment. In order to prevent the low lubricity, the introduction of special additives is required
	Environmental impact. The high penalties are provided for pollution of the environment by vessels. To bring ecological emissions to the required level (without fuel replacement), it is necessary to use various technological devices for the exhaust gases purification and/or special additives, which increases the cost of the vessel construction and fuel costs
	Storage and preparation conditions. They influence the equipment set of the fuel system. There can be special requirements for the materials and coatings, which may increase the vessel construction cost
	Industrial production output can affect the price of fuel and the possibility of bunkering
	Price. It is better to recalculate the price of fuel per unit of combustion temperature rather than volume or mass, which gives a real idea of the fuel specific cost
Ecological	*Density.* When high-density fuels are used, the temperature of the exhaust gases can increase, so the intensive formation of nitrogen oxides can happen as a consequence
	Viscosity. The high viscosity of the fuel in the input of the high-pressure fuel pump can lead to the temperature increase of the exhaust gases and incomplete combustion of fuel (the solids concentration increase in the exhaust gases as a result)
	Sulphur content. Burning of fuels with high sulfur content leads to the formation of a significant amount of sulfur oxides in the combustion products of the fuel
	Ash content. Combustion of fuels with high ash content leads to an increase in the concentration of solid particles in the exhaust gases of the engine
	Vanadium and sodium content. These metal compounds in the fuel intensify the formation of sulfur oxides and, as a consequence of high-temperature corrosion, nitrogen oxides

(continued)

Table 2.2 (continued)

SPP parameters	Degree of impact of the fuels' parameters
	Carbon content. Increased carbon content leads to the deterioration of the exhaust characteristics (temperature rise of the exhaust gases and smoke)
	Water content. Water in fuel is finely dispersed (WFE) and leads to an improvement in exhaust characteristics due to a decrease of the exhaust gases temperature
	Asphaltenes content. The increased content of asphaltenes leads to the exhaust characteristics deterioration (rise in the temperature of the exhaust gases and smoke), and to an increase of the higher aromatic hydrocarbons concentration in the exhaust gases
	Stoichiometric amount of air required for complete fuel combustion. Reduction of the excess air factor leads to incomplete combustion of fuel, and as a result, to an increase in the concentration of fuel non-complete combustion products in the exhaust gases
	Peak flame temperature. The high temperature of the flame can cause intense formation of nitrogen oxides
	Environmental impact. It is characterized by the content of harmful substances in the waste gases during fuel combustion, the duration of decomposition in water, soil or air and influence on living organisms in case of leakage
Flexibility	*Density.* The change in the mass cyclic fuel supply under the operation applying fuels with different densities. The high-density fuels have poor flammability and low combustion rates
	Viscosity. The deterioration of the fuel jet spray, the increase in the duration of engine start-up and transition from one mode to another
	Pour point. In terms of operation during the cold season, the fuels with a high pour point increase density and viscosity, which leads to a malfunction of the engine
	Filter clogging point. The filter clogging and formation of plugs in the fuel-wires can lead to a decrease in the mass cyclic fuel supply and problems with the engine start
	Carbon content. Increased carbon content leads to ignition deterioration and fuel combustion slowdown
	Lower heating value. Transition to a fuel with a lower heat of combustion requires re-tuning of the fuel cells in order to increase the mass cyclic fuel supply to ensure the nominal engine power
	Cetane number. It affects the autoignition, fuel combustion quality and rate. When the cetane number decreases, these characteristics deteriorate and thus there can occur problems with the engine start
	Acid number can influence the cetane number
	Stability. Instability of fuels can lead to the formation of precipitation, and, as a consequence, to filters and separators clogging and the fuel supply system operation failure

(continued)

Table 2.2 (continued)

SPP parameters	Degree of impact of the fuels' parameters
Reliability	*Density.* Increase in carbon formation and wear of the CPG parts can be the result of incomplete combustion of fuels with high density
	Viscosity. Both low and high viscosity of fuels can lead to a number of problems: increased wear of fuel cells, emergence of cracks in the HPFP
	Pour point (clouding). The probability of the fuel solidification in the double bottom tanks and loss of fuel supply
	Filter clogging point. Frozen fuel can clog filters and fuel lines, resulting in engine stops
	Sulphur content. Combustion of fuels with high sulfur content causes a number of negative phenomena: deposition of carbon, oil film deterioration, formation of resins in fuel, corrosion wear of engine parts and fuel cells, low-temperature sulfur corrosion
	Ash content. The ash causes abrasive wear of engine parts and fuel cells; ash deposits lead to intense high-temperature corrosion
Reliability	*Vanadium and sodium content.* Compounds of these metals in the fuel intensify the formation of sulfur oxides, and it leads to the CPG low-temperature corrosion, high-temperature corrosion, overheating and burn-out
	Carbon content. The high content of carbon leads to the growth of soot deposits, and, as a result, to autoignition deterioration and combustion slow-down
	Aluminosilicates content. The high content of aluminosilicates in fuel causes intense abrasive wear
	Mechanical impurities content. It causes abrasive wear of fuel cells and parts of the CPG
	Water content. Water promotes the development of the fuel cells corrosion and sodium-vanadium corrosion
	Asphaltenes content. The increased content of asphaltenes results in sludging and sedimentation, burnout of the piston head metal, combustion of the protective oil layer on the cylinder mirror
	Stability. Instability of fuels can lead to sludge accumulation, clogging of filters, separators, disruption of the fuel supply system
	Cetane number. During combustion of fuels with the low cetane number, the process becomes less smooth, shock loads increase
	Acid number influences its compatibility with various materials and corrosive aggressiveness
	Copper-plate test. The fuels which failed the test cause corrosion of copper parts
	Lubricity. The use of fuels with low lubricating properties causes intense abrasive wear
	Propensity to microbiological contamination. Under the influence of the products of the microorganisms' vital activity, the metals' corrosion and corrosion of fuel tanks walls occurs
	Peak flame temperature. The occurrence of the high-temperature corrosion at high flame temperatures

(continued)

Table 2.2 (continued)

SPP parameters	Degree of impact of the fuels' parameters
	Storage conditions affect the service life and overhaul period of fuel system elements

On average, the emissions of aldehydes while the engine operates on alcohols are about 2–4 times higher than that on gasoline. A significant disadvantage of this type of fuel is its high cost (1.8–3.7 times more expensive than oil).

In addition, alcohols are hygroscopic, have poor lubricating properties, are corrosive (due to oxidation to the corresponding acids), poorly combined with structural materials. Their direct use requires certain changes in the design of the engine. Cetane numbers of alcohols are very low, and serious difficulties in applying them to diesel engines are associated with this fact [7].

Methanol and *ethanol* are the most widely used as motor fuels, and they have the calorific value two times lower compared to petroleum fuels, which means doubled consumption to provide the same engine power.

Methanol is a pure liquid alcohol that can be obtained from natural gas, coal, crude oil and biomass, such as wood residues, directly by catalytic synthesis. For the production of methanol the synthesis gas is mainly used, but if it is necessary to produce large volumes, the preferred raw material is natural gas. Currently, methanol as a motor fuel is used in limited quantities, although in Brazil it is widely used in transport. It is mainly used for the production of synthetic liquid fuels, as a high-octane fuel additive or as a raw material for the production of the anti-knock compound called methyl-tertiary-butyl ether.

One of the most serious problems of the methanol compounds use is the low stability of gasoline-methanol mixtures and their sensitivity to water. The difference between the gasoline and methanol densities and high water solubility of methanol lead to the fact that even small amounts of water within the mixture cause immediate layering, and the layering tendency increases with the temperature decreasing, water concentration rise, and the content of aromatic compounds in gasoline reduction. To stabilize gasoline-methanol mixtures, the following additives are used: propanol, isopropanol, isobutanol and other alcohols.

The pure methanol can be used in specially designed engines, for example, for some race cars. Its high octane number is very effectively realized with account for high compression engines that develop significantly more power than similar gasoline engines. Methanol can be mixed with gasoline for use in engines that can run on several fuels.

Although the emission of CO, nitrogen oxides and hydrocarbons in specialized methanol cars is lower, the exhaust of such transport contains more formaldehyde. When methanol burns, no solid particles or sulfur oxides are released, and it releases less nitrogen oxides than any other fuel. The cost of methanol compared to gasoline is high. Methanol is extremely toxic, therefore dangerous while transported; it causes corrosion of parts in contact with it, which requires changes in the fuel system of the engines. The calorific value of methanol is two times lower than that of gasoline (16 and 32 MJ/l, respectively), as a result, the fuel consumption per unit volume increases and the time between refueling decreases, which is compensated by some stability range for use at high compression ratios and the ability to provide more power [8, 9].

Ethanol as an additive to fuels is more effective than methanol, because it is better soluble in hydrocarbons and less hygroscopic. The phase stability of the ethanol-fuel mixtures is higher than that of the methanol-fuel mixtures, but they also require stabilization. Ethanol can be used directly as a fuel, but, like methanol, it has higher evaporation heat than gasoline, so the engine cold start is a problem that is removed when the mixture of gasoline with 20% of this alcohol is applied. Ethanol is less toxic and less corrosive than methanol, although their technical characteristics and emission levels are almost the same. A positive aspect for the environment is that ethanol is a renewable source of energy, and in some cases it can be produced even from waste. However, ethanol has also disadvantages, such as the following:

- ethanol as alcohol contains the hydroxyl group. If the required amount of water is not removed from the gasoline-ethanol mixture, gasoline can float upward, ethanol will break up, which will lead to a breakdown in the stability of the mixture. At the same time, ethanol is better biodegradable or decomposed into nontoxic components than gasoline;
- ethanol is produced from agricultural crops, thus, large areas are required for its production;
- along with the reduction of CO emission with alcohol fuels, the release of aldehydes increases, which influences the eyes badly;
- as for methanol, the potential level of greenhouse gas emissions depends on the feedstock and the method of production. The full cycle of greenhouse gas emissions for ethanol is 70–80% of the level characteristic of gasoline in the production of corn and 10–100% in the production of wood. The emission of CO_2 during combustion of only alcohol is the same as for gasoline at the corresponding energy equivalent.

The seriousness of the attitude in the world to alcohol fuels is determined by the level of their application in vehicles. In the 90 s, in Brazil, pure ethanol was used by more than 7 million cars as motor fuel, and its mixture with gasoline (gasohol) was also used by more than 9 million cars. In this country, ethanol is made from sugar cane and a car can be refueled with 25,000 ethanol refills. However, in recent years there has been a decline in the use of ethanol as motor fuel.

The United States is the second world leader in the scale of ethanol production for motor vehicles. Ethanol is used as the "pure" fuel in 21 states; the ethanol-gasoline mixture accounts for 10% of the US fuel market and is used by more than 100 million engines [8, 10].

Ethers are better soluble in fuels, less hygroscopic and less corrosive than alcohols. Ethers are traditionally added to automotive fuels. In recent years, interest in dimethoxymethane, dimethyl and diethyl ethers as components of diesel fuel has become apparent. To a large extent, this is due to their good ignitability in the engine and, consequently, high cetane numbers.

Dimethyl ether can be directly injected into the combustion chamber of the engine or used as an additive to liquefied gas, methanol or standard diesel fuel (DF in general, MGO—marine gas fuel—when talking about marine application). A special fuel supply system is required for dimethyl ether injecting, which is a gas under ordinary conditions, since this ether has low lubricity, low viscosity and, like all gases, it is easily compressed. When dimethyl ether is used as an additive to the base fuel, the injection problem is simplified, and other problems are solved at the same time. For example, dimethyl ether increases the cetane number of methanol. As shown by the tests, when engines operate on dimethyl ether, soot formation is almost absent, however, the emission of nitrogen oxides increases, which requires the engines equipment with catalytic neutralizers [9, 11].

Diethyl ether is even more convenient to use and effective, it is liquid (although low-boiling) and its cetane number exceeds 125 (according to some reports it reaches 160). The addition of 10% diethyl ether to diesel fuel increases its cetane number by 4 units in average, which makes it possible to avoid the use of toxic and explosive alkyl nitrates [9, 11].

Hydrogen can be used not only in internal combustion engines, but also in fuel cells that produce electricity. According to the European Commission project, developing a program for introducing alternative fuels, by 2020 hydrogen will be accounted for at least 5% of the total consumption of motor fuels [12, 13].

Methane is called ***shale gas***, which is found in highly clay-dense tough dense rocks: silts, mudstones and shales. Unlike natural gas, it is not concentrated in underground traps, but is distributed at large depths in the pores of the rock, which considerably complicates its extraction and increases its cost. The release of such gas on the surface is prevented by clay layers and denser rocks lying above [14]. The density and calorific value of shale methane is two times lower than that of conventional gas. The most active exploration and production of shale gas is in the United States, where such gas producers are provided with privileges and subsidies from the government [15].

Coal origin fuels. Synthetic oil and chemical products can be obtained from coals. As a result of a number of successful technological solutions, the direct liquefaction of coal has become quite effective. This caused an increased interest in plastic waste co-processing with coal, especially in the US, Japan, and Europe. Since most plastic compared to coal is rich in hydrogen (up to 14% of mass), it should be expected to be hydrogen source when co-liquefied with coal to get the "coal oil". The waste plastic is a cheaper raw material than coal, but its storage is expensive, so recycling plastic waste into motor fuels and chemical raw material is a serious alternative, taking into consideration the economic and environmental reasons [16].

The main disadvantages of the known technologies of chemical coal processing in comparison with the technologies of oil refining and petrochemistry are relatively low productivity and harsh conditions of their implementation (high temperatures and pressures). To eliminate these downsides in coal processing, catalysts and new catalytic processes are increasingly being used, which make it possible to obtain a variety of fuels and chemical products from coal [17, 18].

Among the new coal technologies, a great interest is provoked by the technology of water-coal fuel (WCF), which arose in the 1950-1960s in the coal hydrotransport. The need to burn watered coal fines led to the development of water-coal slurry (WCS) and methods for their combustion. Further improvement of the technology (improvement of the rheological characteristics of the WCS and its stability by using the results of research in the field of coal chemistry and the additives development) led to the creation of water-coal fuel.

Biofuels. The main areas of biomass use in the energy industry include:

- manufacture of pallets (cylindrical briquettes) and wood chips for direct combustion;
- production of synthesis gas (biosingas, syngas) and biomethanol for transport needs;
- production of bioethanol (alcohol), biodiesel fuel, biohydrogen and biogas;
- production of liquid fuel by means of rapid pyrolysis technologies (pyrolytic fuel).

Bioethanol is a product of microbiological processing of corn and other types of starch according to the scheme "starch—glucofructose—ethanol". Bioethanol can be used as a 10–15% additive to gasoline during production of the fuel elements (ethyl tert-butyl ether, biodiesel) and as chemical raw material (in the production of ethylene, butadiene).

Liquid fuel was first obtained in Canada from biomass. The liquid fuel is the product of thermal processing of wood and plant residues, followed by the capture of pyrolysis components by special cyclones. In Ukraine, in the scientific and production complex of gas turbine construction "Zorya-Mashproekt" together with the Canadian company "Orenda", several studies of biofuels from biomass were carried out, which confirmed the possibility of its use onboard of ships and land gas turbine engines, both of the domestic production and engines of the "Orenda" company without constructive refinement [4].

Pyrolysis is usually used in the processing of lignocellulosic materials without air access to produce liquid organic fuels. Rapid pyrolysis technology offers an efficient way to produce liquid fuels from biomass, which is one of the cheapest today, according to the International Energy Agency (IEA). In the conditions of constant growth of prices for oil and oil products, as well as a shortage of traditional energy carriers, the production and use of liquid fuels from biomass can mitigate energy tension in individual countries and increase their energy security.

The liquid formed during pyrolysis is called "*pyrofuel*", "pyrolysis liquid", "pyrolysis oil". Untreated pyrofuel is a thick black resinous liquid, the output of which can reach 70% of the dry weight of raw materials. It is close in its composition to biomass, has higher combustion temperature and consists of a complex mixture of highly oxidized hydrocarbons with substantial water content (Table 2.3).

Pyrofuel has much higher energy density than the raw materials do, which gives it an advantage in transportation and storage. It can be used as a substitute for boiler

Table 2.3 Untreated pyrofuel parameters

Parameter	Value
Water content, %	15…30
pH	1.5…3
Density, kg/m^3	1200
Superior calorific value, MJ/kg	16…21
Viscosity (under 40 °C and humidity of 25%), centipoise	40…100

                                             a   *b*

Fig. 2.1 Complex of solid domestic wastes utilization according to the MCP technology: **a**—version 1; **b**—version 2

fuel, there is a positive experience of its use in gas turbines and diesel engines. After further processing, the pyrofuel can replace motor fuel [19].

Biogas is one of the most promising types of motor fuels produced from local raw materials, from the point of view of industrial production and the vehicles' engines use. It consists of methane (55–75%) and carbon dioxide (25–45%), and the combustion temperature depends on the methane concentration and is 21–29 MJ/kg.

In the Admiral Makarov National University of Shipbuilding under the leadership of the President, Doctor of Technical Sciences, Professor S.S. Ryzhkov, the technology of multi-circuit circulatory pyrolysis (MCP) for obtaining *the liquid fuel of light fractions from domestic solid waste*, including medical ones, has been developed. The uniqueness of this technology lies in the ability to control the level of thermal destruction of the entire mixture of high-molecular toxic components, which can become less toxic with a decrease in their molecular weight. It does not demand deep purification of smoke gases, and the resulting products meet all environmental requirements. This technology at the deepest degree of decomposition does not involve the use of intermediate catalysts, which is significantly different from the other similar technologies. Capital costs for the construction of such processing plant are ten times lower than those of foreign analogues, while operating costs are by hundreds of times lower, as there are no expensive catalysts for flue gas purification. Figure 2.1 shows the versions of circuit solutions of the complex using the MCP technology [20].

2.3 Perspective Alternative Fuels for Application on Marine Transport

Alternative fuels (AF) for SPP are any fuels, except for the fuels of light and heavy oil origin, which can be burned in thermal engines or boilers, or used as a source of hydrogen for fuel cells. To make a decision on the possibility of using an alternative fuel for the SPP operation, it is necessary to meet its following requirements [5]:

- *physical and chemical parameters* (composition, viscosity, calorific value, cetane number, impurities content, etc.): the closer these AT parameters are to the traditional fuels parameters used in the SPP, the fewer modifications are required for the engine and fuel system;
- *bunkering organization*: the availability of raw materials for fuel production, the global volume of industrial production of fuel, the distribution of enterprises producing fuel around the world, the organization of delivery to ports, the receiving equipment of ports, fuel storage and providing of vessels bunkering;
- *ship fuel storage conditions*: the composition and mass-dimensions of the fuel system and equipment, the requirements for maintenance and repair, fire and explosion safety, restrictions on the use of certain types of vessels and heat and power equipment;
- *ecological characteristics*: the composition of the combustion products, the impact on the environment and human health in case of leaks or evaporation;
- *economic effect*: direct costs of fuel, modernization of existing or the new fuel system equipment design, cost of maintenance and repairs, overhaul operation time and equipment durability, share of fuel costs in total operating costs per vessel and power plant.

The AF analysis attracts the most interest. These AF are already used on ships, or their prospects have been confirmed by experimental tests.

Nowadays coal, formerly the main fuel for ships, is practically not used in the SPP. First of all, this is due to the fact that when it is burned and processed, more harmful co-products are formed than during oil and gas combustion. Reduction of damage to the environment from coal energy can be achieved by moving to the use of environmentally friendly fuels of coal origin. These include so-called "pure" coal, water-coal slurry, synthetic gaseous and liquid fuels obtained by chemical processing of coal. Emissions of harmful substances using these synthetic fuels are significantly lower compared to the use of conventional coal [17].

A number of recent publications indicate a renewed interest in *coal fuel* usage in ship power engineering. Provision is made for the coal combustion in specially designed boilers of steam turbine power plants. By adding special inert materials during combustion, the levels of nitrogen oxides and sulfur in the EG are reduced, regardless of the grade of coal burned. Serious attention is paid to the complex fuel preparation system, which necessarily includes a fuel grinding system [21].

The following fact is distinctive in this regard. Onboard of the "Rikes Boin" with 75,000-ton deadweight, the Japanese company Mitsubishi installed two coal-fired steam boilers. Steam boilers with a capacity of 70 t/h of steam superheated to 480 °C at a pressure of 6.3 MPa were equipped with an automated feeding system for granulated (22…45 mm) coal.

The coal was stored in storage bunkers with a volume of about 4000 m^3, from which it was supplied into two supply hoppers (hoppers), designed for six hours of operation of steam boilers. From the hoppers, the pneumatic system supplied 16 t/h of coal to the furnaces of the boilers. Removal of soot from the storage tank with a capacity of 25 tons was produced by water pulp. All systems for the coal receiving, supplying and storing, soot removing were tight, which eliminated the appearance of dust in the boiler room. The power plant with a 14-MW steam turbine unit MS-21-2 with a specific coal consumption of 660 g/(kW·h) provided the vessel with a speed of 16 knots [22].

Liquefied petroleum gas (LPG) by 90–95% is a mixture of propane and butane with an admixture of heavier hydrocarbons. According to the provided power and environmental characteristics of engines, liquefied petroleum gas is close to liquefied natural gas.

Traditionally, liquefied petroleum gas is used only for land transport. There was a project on the use of the LPG for vessels in the Caspian Shipping Company in Azerbaijan in the 1980s, in which diesel engines ran on a mixture of diesel fuel and oil gas, and the use of LPG on ferries and gas carriers of the Caspian fleet was provided. The motivation for such studies was the possibility of applying bunkering of this fuel within the ports of the Caspian Sea, since a significant amount of associated petroleum gas was formed in oil production at local offshore oil fields [23]. Along with this, LPG carriers can burn the gas in engines, evaporating it during transportation.

In 2005, the project "Green Fish" was launched in Spain. The aim of the project is to reduce fuel consumption and improve the ecological characteristics of fishing vessels. The company-manufacturer of engines "Guascor" has developed two models of marine diesel engines: the bi-fueled SDF180TA-SP, which can operate on a mixture of oil gas and diesel fuel in different proportions (rotation speed is 1500 rpm, power with fuel mixture is 245 kW, with purified diesel fuel—370 kW), and FPLD180-SP, which uses only LPG (power—220 kW). The experience of using these engines has shown a reduction in fuel costs, because oil gas is cheaper than diesel fuel. Improvements in the environmental performance of engines were noted [24].

Hydrogen can also be used in internal combustion engines and fuel cells (FC). Onboard of the vessel, the gas can be stored as part of metal hydrides (chemical compounds of hydrogen and metals). According to the project of the European Commission, which is developing a program for the introduction of alternative fuels, until 2020 hydrogen will be accounted for at least 5% of the total consumption of motor fuels. In Turkey, a project to create passenger catamarans with internal combustion engines working on hydrogen has been started. The Hydrogen Engineering Center (USA) is developing the design of such engines [25].

Fuel cells are electrochemical devices, the generation of electricity in which occurs directly onboard of the vessel due to the process of reverse electrolysis, i.e., water and electric current are formed during the reaction of hydrogen and oxygen. Compressed hydrogen or methanol is used as the hydrogen-containing fuel. Reactants enter the FC and the react products come out, so the reaction can proceed as long as reagents are available and the performance of the element itself remains the same. Hydrogen fuel cells can generate electrical power for the propulsion motor onboard of the vessel, thereby replacing the internal combustion engine, or used to provide onboard power. There are several types of FC: solid oxide, proton exchange membrane, reversible, direct methanol, melt carbonate, phosphoric acid, alkaline. At the moment, the cost of fuel cells is very high [26].

The consortium FellowSHIP (Fuel Cells for Low Emissions Ships) was established to introduce hydrogen fuel cells into maritime transport in Europe. The Fellow-SHIP consortium includes the classification society Det Norske Veritas (DNV), Eidesvik Offshore, MTU CFC Solutions, Vik-Sandvik and Wärtsilä Automation Norway.

In Europe, there are also established and are functioning:

- the consortium Fuel Cell Boat BV, which includes the following companies: "Alewijnse", "Integral", "Linde Gas", "Marine Service North and Lovers";
- the non-profit association of Hydrogen and Fuel Cells onboard of the marine transport (Marine Hydrogen & Fuel Cell Association—MHFCA), which includes 120 organizations.

With the financial support of Wärtsilä under the European Union, a project to create fuel cells based on methanol for commercial ships (METHAPU) was launched, which aims to use alcohols (primarily methanol) as marine fuel. Fuel cells based on solid oxides with a capacity of 20 kW have already been created, and soon those with 50 and 250 kW will be developed, too [27].

Figure 2.2 presents the diagram of the Solid Oxide Fuel Cell System (Solid Oxide Fuel Cell System—SOFC) developed within the framework of the METHAPU project [27].

The system can be used for combined production of electricity and heat. It can be used onboard of the vessel as part of the auxiliary power plant. The cell consists of three main parts: anode, cathode and electrolyte between them.

Air enters the cathode during its operation and decomposes to form an oxygen ion (O^{-2}). At the same time, gaseous fuel enters the anode of the fuel cell. Fuel should be in the form that can be used in the cell. Preliminary fuel is supplied to the primary converters. Natural gas, which is not completely decomposed at the anode, is returned through the anode recirculator to the primary converters. Any fuel is converted to hydrogen and carbon monoxide inside or outside the fuel cell.

The oxygen ion obtained at the cathode passes through the electrolyte by means of ion exchange, connecting the hydrogen and carbon monoxide. In this case, water

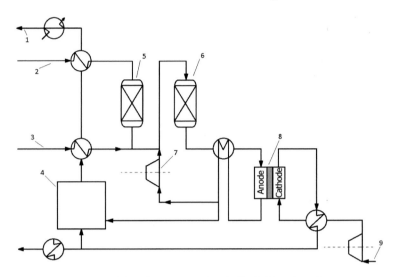

Fig. 2.2 Diagram of the Fuel Cell System Based on Solid Oxides: 1—exhaust gas output; 2—natural gas supply; 3—water supply; 4—catalytic burner; 5, 6—primary converters of gaseous fuel; 7—anode recirculator; 8—solid oxide fuel cell; 9—air supply to the cathode

and carbon dioxide are formed. During oxidation two ions are released, which enter the cathode through an external electric circuit. The process temperature is 650–1000 °C, i.e. there is the possibility of heat recovery. The SOFC can have different physical properties, depending on the shape—flat or tubular.

The main advantages of the FC on ships are its environmental friendliness, high efficiency and compact dimensions (fuel cells have lower mass and take up less volume than traditional power sources, produce less noise and less heat).

The following problems can be considered as the FC application disadvantages:

– the majority of elements produce a significant amount of heat during their operation, which requires the creation of complex technical devices for FC utilization. However, the high temperature of the process makes it possible to produce thermal energy, which significantly increases the efficiency of the power plant;
– there is a problem of obtaining and storing hydrogen. It should be pure enough to eliminate rapid poisoning of the catalyst, and cheap enough to make the cost profitable for the end user;
– the introduction of fuel cells in transport is prevented by the lack of hydrogen infrastructure and a higher cost of energy.

Water-fuel emulsions (WFE) can be produced on the basis of gasoline, diesel fuel, fuel oil residue and other liquid fuels. The interest in WFE is primarily based on diesel fuel and fuel oil residue. Fuel emulsions can be obtained by different technologies: mixing by high-speed agitators, vibrocavitation, ultrasonic, and others [28–31].

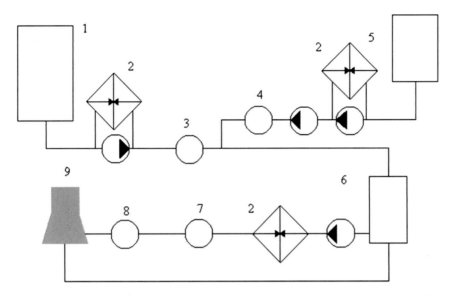

Fig. 2.3 Scheme of the water-fuel emulsion preparation unit: 1—heavy fuel storage tank; 2—preheaters; 3—heavy fuel flowmeter; 4—water flow meter; 5—tank with water; 6—mixing tank; 7—homogenizer; 8—emulsion water content meter; 9—engine

The organization of the engine operation on emulsified fuel requires the inclusion of the technological unit for emulsion preparation in the fuel system of the SPP (Fig. 2.3) [32]. The choice of technology and method for obtaining emulsified fuel (flow or bulk) on the ship depends on many factors and requires detailed analysis. The chosen preparation method should ensure the production of high-quality stable emulsions with a specified dispersion and water content level [28, 29, 31]. The use of WFE on the vessel will increase the productivity of desalination plants and increase operating costs for fuel, which should be taken into account during assessment of the effectiveness.

Combustion of WFE allows improvement of the combustion process quality without significant additional costs. The use of WFE onboard of the ship requires minimal modernization—the fuel systems include equipment for emulsion production and water supply line. The emulsions with water content up to 20% are used for engines and up to 30% for boilers [29].

The use of emulsified fuel has several advantages:

– it reduces carbon formation, increases the reliability of the cylinder-piston group, gas outlet, fuel cells, and the time between the fuel filters purification increases;
– with a decrease in the temperature of the exhaust gases, the thermal tension of the cylinder-piston group parts decreases and the rate of contamination of the lubricating oil also lowers;

– the toxicity and smokiness of the exhaust gases decreases (nitrogen oxide emissions are reduced by 30–50%, carbon monoxide—by 50–80%, sulfur—by 20–60%);
– the possibility of utilization of the oil products contaminated water within the composition of WFE.
– The disadvantages of using the WFE include the following points [29]:
– the need for special equipment for the emulsions preparation;
– low sedimentation resistance (the deficiency is eliminated by using special additives or surfactants);
– the need to increase the freshwater supplies or desalination plants productivity.

Natural gas (NG), consisting primarily of methane, can be used as a motor fuel. Its use in diesel engines is complicated because of the relatively high self-ignition temperature and low cetane number. To overcome these drawbacks, a two-fuel system is used—a small amount of diesel fuel is injected into the combustion chamber as a priming charge, after which compressed natural gas is supplied.

LNG is a cryogenic liquid that is a mixture of hydrocarbons of the series C1…C10 and nitrogen with the methane fraction of 85…99%. The boiling point at atmospheric pressure is $(-162 … -160)$ °C. When transferring LNG to the gaseous state, its properties correspond to the properties of natural gas. Gasification of 1 m^3 of LNG yields about 600 m^3 of natural gas [33].

The disadvantages of using the LNG are the following:

– storage difficulties caused by the need to maintain low temperature in the fuel tanks; before being fed into the engine, the gas must be evaporated and heated;
– carbon monoxide emissions are increased by 6–7% and those of hydrocarbons— by 20–30%;
– increase of thermal loads and decrease of engine reliability;
– higher value of the excess air factor than that for liquid fuels [34].

Advantages of using the LNG as fuel are listed below:

– it can significantly reduce the toxic emissions of combustion products: emissions of nitrogen oxides are reduced by 84–93%, carbon dioxide—by 20–25%, there are practically no solid particles or sulfur oxides;
– the heat of methane combustion is higher than that of oil fuels;
– abrasive wear of fuel cells decreases;
– the cost of LNG is much lower than the cost of the calorific value equivalent amount of fuel oil residue [34, 35].

Biodiesel fuel (BD) has been widely used in diesel engines recently. It is registered by the Environmental Protection Agency (USA) as the environmentally friendly fuel and fuel additive. BD is a mixture of ethers of higher carboxylic acids, which are obtained by the catalytic interetherification of various vegetable oils and fats with methanol or ethanol [36–38]. The characteristics of the product obtained are very close to similar values of diesel fuel from oil. The raw material for the production of biodiesel fuel can be fat, rarely—essential oils of various plants (mainly oilseeds), algae and fats. In Europe, the BD is traditionally produced from rapeseed, in the USA and Canada—from soybean, in Asian countries—from palm oil. At the same time, vegetable oils from other crops, frying oil and fat, animal oil and fish oil are used [37]. The addition of BD to the petrodiesel fuel in the amount of 5–30% is mandatory in several countries of the world. In diesel engines, vegetable oils can also be used directly, but this requires modifications of the engine design [37].

The European Standards Organization has developed standard EN14214 for biodiesel fuel. In addition, there are standards EN590 (or EN590: 2000) and DIN 51606. The first one describes the physical properties of all types of diesel fuel sold in the EU, Iceland, Norway and Switzerland. This standard allows the BD content of 5% within the petrodiesel fuel. In some countries (for example, in France) all diesel fuel contains 5% of biodiesel. DIN 51606 is a German standard designed to be compatible with the engines of the leading automakers, so it is the most demanding one. Most types of biodiesel fuel produced for commercial purposes in the West meet its requirements or even surpass them. In addition to international and national standards of the USA, Australia and other countries, individual leading engine building companies, such as "Caterpillar" and "John Deere", have developed their own requirements for biodiesel fuel [39].

Biodiesel fuel can be used in existing engines without significantly affecting their performance; a standard storage and preparation system is used for it, as well as for diesel fuel. Mixtures of BD with diesel fuel in small proportions (up to 5%) do not affect the characteristics of the fuel system and engine performance. Visual inspection of the high-quality BD should not reveal undissolved water, sediment, slurry and other foreign inclusions. The BD fuel is to be pure, its color may be different, and this characteristic does not affect its quality.

The BD properties vary depending on the feedstock and the technology of fuel obtaining, as well as on the oxygen content. The quality of fuel can be reduced with insufficient depth of raw materials processing. The efficiency of the engine can be reduced by about 10% when it works on BD manufactured with the use of a simplified technology. With regard to biodiesel fuel that does not meet the standards, using the fuel filters fail quickly [40, 41].

Characteristics of the biodiesel fuels derived from various raw materials may differ significantly (Table 2.4). In biodiesel fuel, the unsaturated fatty acids are predominant among vegetable oils, saturated fats and used oils [40, 42–44].

Analysis of the main characteristics of biodiesel fuel based on the data given in [39, 41, 42, 45–52] allows us to establish their features, taking into account their application in the SPP.

1. *The flash point* of B100 is much higher than that of the diesel fuel. This fact indicates that during the production process all excess methanol was removed from the fuel. The presence of residual methanol, even in small amounts, significantly reduces the flash point, adversely affects fuel pumps, elastomers and profile seals, and degrades the quality of the combustion process within the diesel engine. Due to the high flash point, the BD is referred to as a fire-safe fuel.

Table 2.4 Parameters of the different raw materials of biodiesel fuels B100

Raw materials	Lower calorific value, MJ/kg	Cetane number	NO$_x$ Emissions, % (comparing to DF)	Temperature, °C		
				Cloudiness	Filter clogging	Hardening
Soy	39.54	47	15	3	−4	−2
Rape	40.01	56	12	−3	−4	−4
Pork fat	39.77	64	3.5	13	13	11
Beef or lamb fat	39.31	63	2	19	16	14
Tallow	39.08	62	1.8	16	15	10
Used sunflower oil (sample 2)	40.01	58	5.7	–	9	11
Used sunflower oil (sample 2)	39.77	53	2.2	8	6	1

2. **Water and sediments** show the presence in the fuel of unbound water globules and sedimentary particles. The requirements for this characteristic for B100 are the same as for the DF. Watering of biodiesel fuel can be the result of improper transportation, storage, or incomplete removal of water during receipt. The fuel can begin oxidizing, which leads to the formation of sediments. This index, along with the viscosity and acid number, is a criterion for the oxidation of fuel during storage. BD has a high hygroscopicity and actively absorbs moisture, being a favorable environment for the reproduction of microorganisms. This can lead to corrosion of fuel equipment, the emergence of deposits of biological origin in the fuel system, and, as a consequence, to the formation of sludge, blockage of filters and pipelines.

3. **The viscosity** of B100 is slightly higher than viscosity of the DF, which leads to reduction in the fuel equipment leaks. BD is used mainly in diesel engines, the performance of which is determined by the technical state of the fuel equipment. There are data that switching to vegetable-based fuels with a higher viscosity makes it possible to extend the life of the engines even in conditions of extreme wear of the plunger pairs of the fuel pump. At the same time, a higher viscosity of the BD can lead to the deterioration of the combustion process, formation of deposits, spraying of fuel when fed to the engine, getting it into the engine oil. The high density and viscosity of B100 are associated with increase in NO$_x$ emissions. The increase in the nitrogen oxides emissions when the engine is running on a mixture of B20 (a mixture of diesel and biodiesel fuels in the ratio of 80:20) is observed during operation at low speeds, but with the high load or torque. During the long engine operation using the BD fuel, the lacquer deposits

can be formed on the fuel injectors, corrosion and jamming of the internal parts of the fuel injection system may occur, the feed pump may malfunction due to water ingress, formation of sludge and sediment, which leads to decrease in the overhaul period.

4. *The calorific value* of B100 is slightly lower than the calorific value of diesel fuel. Therefore, when the engine is running on biodiesel, the power is reduced.

5. *The sulphate residue* indicates the amount of residual alkaline catalyst in the BD. Its presence in the fuel can lead to sedimentation within the injector and breakdown of the fuel system. *The carbon residue* is the average indicator of the fuel's ability to form carbon deposits within the engine. The carbon residue of DF is determined by distillation of 10% of the residue. For BD it is difficult to determine this value strictly for 10% of the residue, because its boiling point varies in a wide range.

6. For the normal operation of a diesel engine, the fuel must have a *cetane number* of at least 40. A higher value ensures satisfactory engine performance during cold start and reduction in the formation of white smoke. The cetane number of B100 corresponds to a similar indicator of high-quality diesel fuels. BD with a high content of saturated fatty acids can have the cetane number of 70 and higher. To calculate the cetane index of the mixture, it is necessary to know the specific weight and distillation curves of the B100 and DF.

7. *The cloudiness point* for BD is higher than that for diesel fuel. At low temperatures, the fuel loses mobility, becomes gel-like, begins to crystallize, which leads to clogging of filters and pipelines and difficulty in pumping. Some manufacturers report that in the application of additives in the amount of 1% of the fuel mass, the cloudiness point for B100 is reduced by 12 °C. Other tests show that the chemical treatment of fuels with the additive of up to 0.1% reduces the temperature by 3 °C. The work is underway on the development of the winter types of biodiesel fuel by replacing saturated fatty acids with unsaturated fats.

8. *The acid number* of BD is the indicator of the presence of free fat acids within the fuel. Its increased content may be the result of violation of the fuel obtaining technology, or it may indicate the process of fuel decomposition caused by oxidation. Using the fuel with the high acid number can lead to accelerated deposits within the fuel system and reduction of the fuel pumps and filters durability.

9. Incomplete conversion of fats and oils leads to an increased level of total *glycerin*. Insufficient BD purification from the formed glycerin causes an increase in the content of free and total glycerol. The high value of these characteristics leads to deposits within the engine, fuel system and tanks, and can cause filter clogging and other problems.

10. *The phosphorus content* in B100 should not exceed 0.001%, although in some vegetable oils this value is higher. It is necessary to prepare the oils, as its presence can cause failure of catalytic converters of industrial units of the BD production.

11. *The distillation temperature* of 90% of the fuel indicates whether the fuel will react with high-temperature surfaces and substances (for example, with the oil

in the lubrication system). The boiling point of B100 is higher than the one at the acceleration curve and is about (330 … 375) °C under normal conditions.

12. ***The stability of fuel and fuel mixtures*** is a very important performance indicator. There are two stability indicators: stability under long-term storage and normal conditions, and thermal stability (at elevated temperatures and/or pressures when fuel circulates in the engine's fuel system). The loss of stability of the BD due to oxidation or long-term storage can lead to an increase in the acid number, viscosity and formation of sludgy deposits and residue, which leads to clogging of the filters. If the values of the above listed characteristics do not meet the standards, such fuel is not recommended for use, since it will decompose rapidly. B100 with high stability to oxidation will keep its basic performance characteristics unchanged longer. Brass, bronze, copper, lead, tin and zinc can accelerate the oxidation of B100 and lead to the formation of gel-like insoluble deposits at the contact with certain components of the fuel. The available data indicate that B100 has a good thermal stability. This can happen due to the fact that saturated vegetable oils and fats can be used for a relatively long time at high temperatures.

13. ***The purification properties.*** B100 can dissolve deposits (even perennial) formed in the fuel system, equipment, and fuel tanks. Dissolved deposits cause deterioration in the filtering properties of materials, their swelling and fuel leakage; they can clog the filters. The cleaning effect depends on the amount of sediment in the system and the BD concentration at the use of mixed fuels. It increases with the use of B100 and mixtures containing more than 35% of biodiesel.

14. ***The corrosion aggressiveness.*** BD is a chemically and corrosively active liquid. B100 can soften, decompose and cause seepage through sealing materials (in particular, natural rubber, nitrile and synthetic rubber), gaskets, bearings, elastomers, adhesives and plastics during prolonged contact with it, which can lead to leaks through sealing connections and hoses. The fuel inflammation when it gets on the heated engine, fuel pump breakdown and filter clogging can be the result of possible leaks, because the materials incompatible with biodiesel are decomposed and crumbled. The BD can also dissolve some types of paints and coatings during prolonged contact with it. Compounds with nitrile rubber, polypropylene, polyvinyl and polyethylene are very exposed to the impact of B100. Connections made of lead, copper, brass, bronze, and zinc should be protected from contact with the BD. Therefore, the equipment made out of the above materials should be replaced with equipment and fittings made of stainless steel or aluminum. The mixture of B20 is compatible with almost all materials.

15. ***The ecological compatibility.*** When the engine is running on biodiesel, visible smoke and emissions of particulate matter, higher hydrocarbons and CO decrease due to the presence of oxygen in the fuel. This results in the more complete combustion of fuel and reduction in the amount of unburned fuel particles. The BD is neutral with respect to CO_2 emissions, because when fuel

is burned, it emits as much as the plant absorbs in the process of photosynthesis. The release of formaldehyde and nitrogen oxides is somewhat increased.

16. *The toxicity*. The BD is not toxic to living organisms. The tests showed that with a short period of exposure, the fuel poses no threat to humans and animals. There was a slight negative effect on lung tissue with a high level of the exposure dose. 90% of B100 decomposes in 3 weeks in case of leaks or spills, including the cases when it enters the water.

Thus, there are some advantages of using BD fuel in thermal engines [53]:

- it practically does not contain sulfur, so the emissions of sulfur dioxide and sulfur oxides into the atmosphere are significantly reduced;
- it has a high degree of biological decomposition for a relatively short period;
- the smoke gases decrease by two times, while the concentration of CO, hydrocarbons and soot decreases by 25–50%;
- as a product of processing of the plant raw materials, biodiesel fuel does not contain polycyclic aromatic hydrocarbons, especially benzopyrene, and carcinogenic substances;
- due to the higher oxygen content compared to the DF, the BD requires less air for complete combustion;
- due to the high flash point, biodiesel is considered to be fireproof;
- BD improves the lubricity of low-sulfur diesel fuel when using the mixed fuels.

Disadvantages of biodiesel fuel are as follows [53]:

- the low calorific value, which leads to decrease in engine power by 5–16% and increase in fuel consumption;
- the need for frequent replacement of fuel filters and routine maintenance of the nozzles as a result of the sprayers coking;
- the high viscosity, which causes deterioration of spraying, mixture formation and fuel combustion in the engine and is the cause of deposits on the walls of the combustion chamber, as well as rapid engine failure;
- the low stability along with rapid oxidation and tendency to thermal decomposition, which creates a favorable environment for the multiplication of microorganisms;
- the nitrogen oxides emissions increase by 10% compared to DF;
- softening of the thickness made of natural rubber.

Onboard of the ships we can also use non-traditional sources of renewable energy, namely the energy of the sun and wind [54–56]. There has been developed a project of the EOSEAS cruise ship, which includes the SPP operating on liquefied natural gas (LNG) and providing basic energy needs. In addition, the vessel will be equipped with semi-rigid sails with a total area of 20 thousand m^2 [56].

Thus, the most promising fuels alternative to the traditional marine fuels presently are BD, LNG and VTE, which is determined by the following factors:

- these fuels are produced or extracted on the industrially significant scale;
- they are adapted for combustion within ship power equipment (VTE—for internal combustion engines, gas turbine engines and boilers, BD-ICEs, LNG—for gas-diesel engines, GTEs);
- they are non-toxic; as a result of their use, the level of harmful emissions from ships is reduced, and biofuel is renewable;
- their cost is correlated with the cost of traditional fuels, the price of the BD is comparable to that of conventional MGO;
- there is a positive experience of using these fuels onboard of the ships.

2.4 Review of Experience in Alternative Fuels Usage in Ship Power Plants

When analyzing the possibility of using AFs for ships, it is necessary to evaluate their influence on the following elements of the SPP:

- the low-pressure fuel system (the possibility of bunkering, need to modernize existing fuel systems, storage and preparation of fuels, rational equipment parameters);
- the engine (effective performance, environmental parameters, features of the operation process flow);
- the power plant (the SSP parameters and they way they depend on the fuel chosen determine the nature of the performance indicator).

The information on the operating ships and the ships under construction, on which various types of alternative fuels are used, and the analysis of their experience of its use are of great interest.

The fuel cells. The project for the use of fuel cells onboard of the ships has been implemented at participation of the number of European marine and research organizations. The first vessel is a 140-m passenger ferry shuttling between Oslo and Kiel, where high- and low-temperature fuel cells were used instead of diesel generators, diesel engines were part of the propulsion system. As a source of hydrogen for fuel cells, low-sulfur diesel fuel (LNSS), liquefied natural gas and liquefied hydrogen were used. Hydrogen was used in low-temperature fuel cells, diesel fuel and LNG—in high-temperature fuel cells. The installed capacity of the ship power plant was 2 MW. The second vessel was a 30-m recreational craft in Amsterdam, where all energy requirements were provided by fuel cells, compressed water was stored onboard of the ship. Table 2.5 provides the comparison of various technologies for generating electricity onboard of the first of the given vessels [57].

For the second vessel, the efficiency of the ship power plant with standard diesel engines is about 26.8%, and with fuel cells—about 50%. This is due to the fact that on

Table 2.5 Efficiency of various technologies of electricity generation onboard of the ferry

Methods of electricity obtaining onboard of the vessel	Technology efficiency (Performance Coefficient), %
Diesel engines	43.3
Melt and carbonate FC1, hydrogen source—HCDF	41.8
Melt and carbonate or solid oxide FCl, hydrogen source—LNG	47.8
FC with proton exchange membrane, hydrogen is stored in liquefied state	50.0

Table 2.6 Vessel's ecological indicators for various technologies of electricity generation

Pollutant	Emissions level for different methods of electricity generation, g/(kW · h)			
	Diesel engines with heavy fuel	FC with hydrogen obtaining source of		
		Diesel fuel	Liquefied natural gas	Liquefied hydrogen
NO_x	11.8	0.0161	0.0141	0.0068
Solid particles	1.57 (0.37[*])	0	0	0
Volatile organic compound	0.49	0.0079	0.0069	0.0033
CO	1.53	0.0323	0.0282	0.0135
CH_4	–	0.0728	0.0637	0.0304

[*]The data in parentheses are given for heavy fuel with the sulfur content of 1%, without the parentheses—for 3.5%

this vessel the diesel engines operate mainly on partial loads, where the engines are less efficient and fuel cells are more efficient than at full loads. Table 2.6 shows the components of emission factors for substances polluting the air for different methods of the onboard electricity generation [57].

According to the data, the emissions of harmful substances at the FC application are significantly reduced compared to that for heavy fuel. The most so-called "clean" option is the hydrogen storage directly onboard of the vessel, which makes it possible to eliminate additional converters in hydrogen production. The analysis of the environmental efficiency of the production processes of the equipment for implementation of the various technologies of electricity generation also poses some interest (Table 2.7) [57].

The data presented in the tables indicate that polluting emissions during the production process are much higher for fuel cells than those for standard diesel engines. As a result, the total pollution of the atmosphere for the entire "life cycle" of each technology will be comparable.

In 2009, a passenger ship (river tram) for 100 places was put into operation in Amsterdam. The vessel has the fuel cells with power of 60 kW [58]. The world's first vessel utilizing fuel cells as a propulsion unit was put into operation in late 2008.

Table 2.7 Emissions of air pollutants in the equipment production process for various electricity generation technologies

Pollutant	Emission level, g/(kW · h)			
	Diesel engines	Melt and carbonate FC, hydrogen source—HCDF	Solid-oxide FC, hydrogen source—LNG	FC with proton exchange membrane, hydrogen is stored in liquefied state
CO	0.280	1.590	0.374	1.197
NO$_x$	0.260	1.500	0.742	0.679
SO$_x$	0.130	9.200	2.970	1.319
Volatile organic compound	0.030	5.570	0.002	0.236
Solid particles and soot	0.030	1.930	0.647	0.216

The vessel Alstewasser designed for 100 passengers runs on the Alster Lakes and connected channels in Hamburg (Germany). The length of the vessel is 25.5 m, its width is 5.2 m, and its height is 2.62 m. The vessel has two 48 kW electric motors receiving fuel from the fuel cells. Cruise speed of the ship is 15 km/h. The mass of the gaseous hydrogen reserves is 50 kg [59].

Of great interest are the projects of ships where the propulsion unit is equipped with bi-fuel engines powered by natural gas, which is stored onboard in the liquefied form. The demand for electricity is provided partly by fuel cells. Examples of such vessels are the "Viking Lady" service ship and the "Glutra" ferry. The latter is equipped with the diesel-electric SPP with gas-diesel engines and fuel cells with the proton-exchange membrane (Fig. 2.4) [60]. In addition to the fuel cell, the container has a reformer (reactor for processing liquid hydrocarbons into gas) and preliminary air conditioning system that serves as a source of oxygen.

The "Viking Lady" ship is a part of the Fellow SHIP project, developed by the classification society Det Norske Veritas, ship owner Eidesvik, Wärtsilä and MTU Onsite Energy. The power of the fuel cells installed on the "Viking Lady" is 320 kW, and the propulsion diesel-electric power plant contains four Wärtsilä 34DF engines with the capacity of 2010 kW each. As a source of hydrogen for fuel cells, natural gas is used. The future plan are to increase the capacity of the fuel cells installed on the ship to 1–4 MW with practically the same weight and size parameters of the unit [61].

Analyzing the experience of using the WFE for ships, the developed schemes for preparing emulsions for diesel engines under ship conditions, GFE and boilers are of primary interest [29, 31].

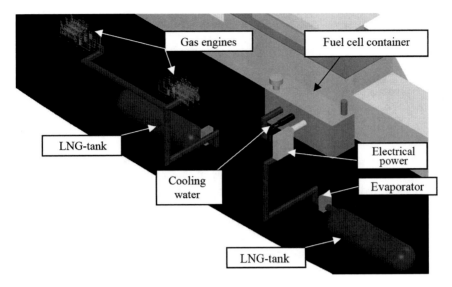

Fig. 2.4 Location of fuel cells onboard of the "Glutra" ferry

During development of these systems, it is necessary to take into account the convenience of maintenance, duration of the emulsion storage, possibility of optimizing the operation of the engine in various modes, grade of base fuel and other factors.

Preparation of the water-fuel emulsions can be carried out at the shore bunker stations (bases) with subsequent transfer to the ship within a special ship unit or directly into the diesel engine. In the first case, problems arise because of the limited kinetic stability of the emulsion, which requires additional equipment to maintain the dispersed medium in suspended state, both onboard of the base and the vessel. Stabilization of the WFE and equalization of its components can be achieved by means of the special surface active agents (SAA) usage. However, additional measures are also required here, such as introduction of metering devices into the system for the WFE preparation, as well as means of the SAA storage and transportation to the unit.

Preparation of the WFE onboard of the ship is possible by two methods: cyclic and continuous. In the cyclic method, the emulsion enters a separate storage tank, from which, if necessary, it is supplied to the diesel engine. This method is used in the episodic operation of a diesel engine with WFE. It has drawbacks typical for the WFE shore preparation. The method of continuous emulsion production assumes its supply directly to high-pressure fuel pumps of the diesel engine.

In the system of the water-fuel emulsions preparation, ***the homogenizers*** can be applied, as they are designed for hydrodynamic fuel processing. The complexity of developing the system for the WFE preparation is in creation of the dispenser that provides the required quantitative composition of the emulsion, depending on the engine operation mode.

The experience of marine diesel engines operation in the West Siberian river shipping (WSRS) made it possible to formulate the basic requirements for the system of the WFE preparation onboard of the ship listed below [28].

1. The simplicity of design, portability and sufficient autonomy, applied elements availability.
2. The automation of mixing and dosing of the WFE components (depending on engine load). The water particle size of the DT fuel is to be up to $(10–12) \cdot 10^{-6}$ m.
3. When using the heavy fuels, the pre-separate heating of fuel and water up to the temperature determined by the viscosity of the emulsion is to be ensured. The device used for emulsification must have the properties of the homogenizer.
4. The fuel and water must be filtered separately, using standard or special filter elements. Water is to satisfy the requirements of sanitary rules for washing water on inland navigation vessels. The emulsion goes through a special filter before being fed into the engine, preventing gelation or breakdown.
5. The design of the WFE preparation system should exclude the possibility of the appearance of stagnant zones and areas where the velocity of the emulsion is lower than permissible.
6. In order to control the WFE preparation unit, a remote system must be provided to ensure the transfer of the engine from the diesel fuel to the emulsion and the reverse operation.
7. The cycle feed of the high-pressure fuel pump (HPFP) should be by 20–30% higher than that at the pure fuel operation.
8. A special measuring device must be provided for continuous monitoring of the quantitative composition of the emulsion within the unit.

The system for the emulsion preparation with the water supply at constant pressure, developed and tested at Newcastle University and implemented onboard of the ships of the Irish company "White Liners", is presented below (Fig. 2.5) [28].

Water and fuel are supplied to the suction cavity of the dispersant, and its discharge pipeline is connected to the fuel line section to the high-pressure feed pump. The constant water pressure is maintained by bypass valve 11, and its flow rate is regulated by needle valve 9. The solenoid valve block is designed for remote control of the system, which allows operation of the emulsion diesel only in the constant mode.

In most cases for the auxiliary ship boilers, the waste oil products are used as the emulsion in the form of contaminated fuels and waste oils with the increased content of water and mechanical impurities that are not subject to use and can be dangerous for the environment. The Riga branch of the CNIIMF offered the fuel preparation system that includes mixing and dispersing of the mixture of the fuel, oil and bilge water separation residues with bunker fuel for auxiliary boilers (Fig. 2.6) [28].

The results of the system operation show that the emulsified liquid fuel burns faster than the anhydrous liquid fuel. The water content of up to 20% in the emulsified fuel not spoils but intensifies the combustion process due to the increase in the evaporation surface of the particles and the intra-furnace crushing of the droplets. Reducing the burning time of emulsified fuel positively affects the afterburning of

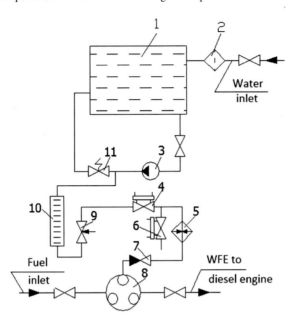

Fig. 2.5 The scheme of the unit for WFE production at constant pressure: 1—charge tank; 2—water filter; 3—circulating water pump; 4, 6—solenoid valves; 5—electric heater; 7—non-return valve; 8—dispersant; 9—needle valve; 10—water flow meter; 11—pipeline valve

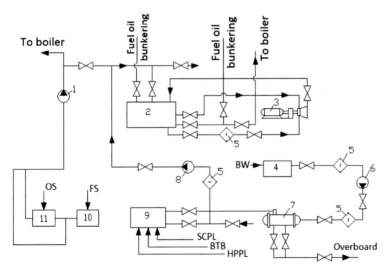

Fig. 2.6 Schematic diagram of the fuel preparation and oil products waste utilization in boilers: 1, 8—sludge and leakage pumps; 2—charge tank; 3—rotor dispersant; 4—bilge water collection tanks; 5—coalescent filter; 6—bilge water pump; 7—bilge water separator; 9—fuel and oil leakage cistern; 10, 11—fuel and oil sludge tanks; SPL—separator pallets leakage; FTSD—fuel tanks sludge discharge; HPPDL—high-pressure pumps diesel leakage; WO—waste oil; OS—from the oil separator; FS—from the fuel separator; BW—bilge water

the grease residues and reduces the carbon deposit on the working surfaces of the boiler.

The information on the use of water fuel emulsions for GTE is very different and fragmented. This can be explained by great caution and, obviously, prejudice of the designers, since the GTE reliability is largely determined by the stability and efficiency of the processes within the combustion chamber. However, the need to ensure the required purity of exhaust gases inevitably puts the GTE makers in a dilemma: to use the WFE and have fuel economy and the relatively clean exhaust gases, or look for other ways to improve the economic and environmental performance of the unit [28].

Biodiesel fuels in many of their physical and chemical characteristics are close to the oil diesel fuel, so the standard equipment of fuel systems and conventional diesel engines can be used at their application onboard of the vessel. At the same time, in order to ensure efficient operation of B100 and their mixtures in the design and operation of the SPP, it is necessary to take into account the specific features of this renewable alternative fuel.

Biodiesel fuel and its mixtures are used to power the engines of the following manufacturers: "Cummins", "Caterpillar", "Wärtsilä", "MAN B&W Diesel", and "John Deere", which have officially announced their intentions to expand the subsequent use of the BD fuel and provide recommendations on the specifics of the engines operation with this fuel [62].

The Government of Canada successfully implemented the BioMer project on the use of B100 and its mixtures with diesel fuels on cruise ships in Quebec [63]. Currently, this is the most ambitious project related to the study of the main parameters of ship equipment using the BD fuel. As the test results showed, during the BD fuel use, the frequency of the fuel filters replacement increased due to their clogging with sediments, which were dissolved upon the contact with this fuel. The phenomenon of filter clogging was observed after 4–8 weeks of the vessel operation on the BD fuel. The testing of the filters for 300 h did not detect any negative influence of B100 on the filter material or other filter components. As a result of the BD fuel having higher viscosity, the pressure drop of approximately 25% was observed (Fig. 2.7).

The filter pressure increased by 58–66% at the inlet and by 44–46% at the outlet; the lower the BD content in the mixture is, the less prominent this effect is. The test on B20 usage showed almost the same results as with DF. The pressure drop can be compensated by the use of filter materials with lower density, installation of two parallel filters or a booster pump [63].

The rate of decomposition is an important ecotoxicological indicator determined by the oxygen content of the fuel: the faster the fuel is oxidized, the safer it is. Test results within the BioMer project showed that B100 decomposes by 2.5 times faster than DF does (Fig. 2.8). So, over 8 days at the temperature of 20 °C, B100 decomposes by 80%, B20—by 49%, and DF—by 41%. The rate of decomposition increases with the addition of special nutrients, so that microorganisms multiply much faster [63].

Natural gas on ships is mainly used in liquefied form, which is associated with the requirements for fire safety and volume limitation to place the fuel tanks [64],

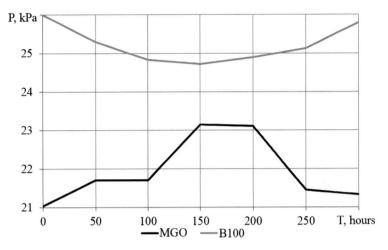

Fig. 2.7 Change in the pressure drop of fuel filters when using DF and B100 over time

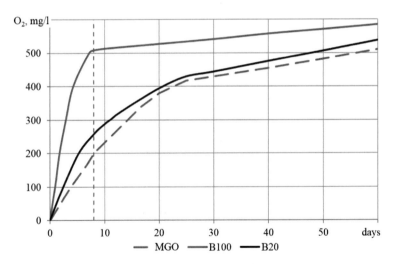

Fig. 2.8 Change in oxidation stability (Oxygen Content) of fuels at their storage

although there is experience of the SPP operating on compressed gas at the pressure of 20–25 MPa [65]. Since LNG is physically, chemically and operationally different from marine oil fuels, its use onboard of the ship leads to a drastic change in the fuel system of the SPP, the engines, and, sometimes, in the type of the propulsion system.

NG as a fuel has been traditionally used and is currently used on gas carriers (evaporation from cargo tanks), storage vessels, production and unloading vessels for the gas production platforms service, which are practically floating plants liquefying the natural gas [66]. Over recent 5–10 years, the ship power plants that do not transport it have started operating on LNG. The composition of the fuel system for LNG will be significantly different [67].

Coast guard vessels, vessels for gas production platforms, transportation and passenger ferries, passenger ships, Ro-Ro and Ro-Pax vessels, catamarans, tugboats, cruise ships, and multifunctional cargo ships all can operate on LNG [67–83].

The DNV Classification Society was the first to develop the rules for the gas use onboard of such vessels, as Norway's gas fleet is presently the largest (referring to the ships where LNG is not transported but only used as fuel) [72]. This is due to the fact that Norway has its own substantial reserves of natural gas, which eliminates problems with bunkering. In addition, a part of the coast of this country is washed by the North Sea, being not far from the Baltic Sea, and the waters of both seas belong to the SECAs (Sulphur Emission Control Areas). In Norway, there is a project on the use of biogas on ships, and this biogas will also be stored in liquefied form [80].

Since diesel engines dominate in ship power plants, the analysis of the experience of using alternative fuels in engines of this type is of interest.

The WFE usage in marine diesel engines is sufficiently considered in the current paper [31]. Let us consider some features of the working process of the diesel engines operating on WFE. Switching to water-fuel emulsions improves the basic parameters of the mixture formation. At the WFE burning within the combustion chamber, there occur microexplosions that break asphalt-resinous compounds. The remains of the emulsion drop reduce the size of the formed carbon particles, which enables their faster combustion in contact with the oxidizer and acceleration of the gasification process. This increases the local coefficients of excess air, which leads to the increase in the combustion rate and reduction in soot formation. To ensure that the engine power during the transition to the emulsion remains unchanged, it is necessary to increase the cyclic feed rate.

At the operation on WFE, it is possible to increase the fuel injection advance angle, which is associated with lower compressibility as compared to petroleum fuels (diesel and motor). At the same temperatures, the viscosity of fuel emulsions is higher than that of pure fuels, which causes a slight increase in the injection pressure. The transition of diesel engines to WFE is accompanied by a significant reduction in the smokiness of the exhaust gases. Along with this, there has been a decrease in the nitrogen oxides emissions, due to the fact that the local temperatures of the working fluid in the combustion chamber decrease during diesel engines switching to emulsified fuels [31].

Natural gas can be used both for bi-fuel engines, which operate on liquid petroleum fuel and natural gas, and for gas engines consuming only gas. Ship internal combustion engines that can operate on the NG can be divided into three groups [76, 84]:

- gas-diesel (GD) engines, which can run on liquid fuel or natural gas with the ignited DF fraction of about 5% (in both cases the engine runs in the Diesel Cycle);
- gas (spark-ignition gas-SG) engines, which operate only on natural gas in the Otto Cycle;

Table 2.8 Emissions of harmful substances in the ICE exhaust gases

Engine type	Diesel			Gas-diesel (Natural gas operation mode)		
Load, %	50	75	100	50	75	100
Harmful substances emissions, g/(kW·h)						
NO_x	11.5	122	12.9	1.6	1.5	1.4
CO	11	0.8	0.7	7.1	5.4	4.5
Hydrocarbons	0.5	0.7	0.8	16.3	14.9	13.6
Solid particles	0.4	0.3	0.3	0.1	0.1	0.1
Volatile organic wastes	*	*	*	2.0	1.8	1.6

*no data available

– dual-fuel (DF) engines, which can run on natural gas with the ignited diesel fuel fraction of about 1% in the Otto Cycle, or on oil fuel in the Diesel Cycle.

These engines can run on natural gas with the methane number of at least 80, on heavy and light fuels and, upon the application of individual engine manufacturers, even on crude oil [85]. In any case, the ship should provide at least two fuel systems— for LNG and DF, and sometimes the third one—for the heavy fuels.

Table 2.8 shows the comparative data on the ecological characteristics of the diesel and gas-diesel engines [65].

The leading manufacturers of gas-diesel and bi-fuel engines in marine design are "MAN B&W Diesel A/S" and "Wärtsilä", the gas diesel engines are also produced by the Kolomensky Plant, Pervomayskdieselmash, OJSC RUMO, Yaroslavl Motor Plant, "Mitsubishi Heavy Industries Ltd", "Perkins Engines Company Ltd", "Anglo-BelgianCorporation N.V." and other companies [84, 86–92].

Nowadays there is a large number of publications on the testing of engines of different models on *biodiesel fuel from various raw materials and their mixtures* in different ratios with diesel.

Section 2.3 provides information on the BioMer project implemented in Quebec, Canada. The characteristics of the engines were also defined in this project (Fig. 2.9). Figure 2.9 *a* shows the results of the marine diesel engine CAT 3176C testing on DF and B100 made of frying oil, and mixtures.

During the tests, the slight increase in power was observed for both mixtures and for pure BD fuel on average by 2–8%. At 1500 rpm, the increase in the effective power of N_e by 15% was noted when using B5. For all the engine operation modes, the performance improved. Compared to diesel fuel, the efficiency increased by 2.3% for B5 and B20, and by 3.3% for B100. At the same time, the calorific value is lower by 0.3%, 1.4% and 7.2% for B5, B20 and B100, respectively. The fuel consumption (in l/min) decreased by 1.8% and 0.8% for B5 and B20, increased by 3.3% for pure biodiesel fuel [63].

At the same time, according to the results of the diesel engines tests presented in other works, there is a decrease in engine power when operating on B100 on average

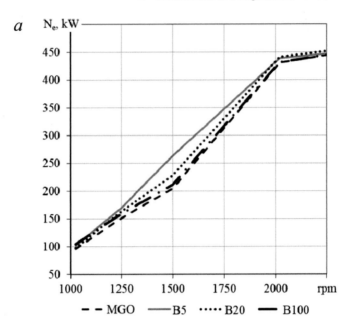

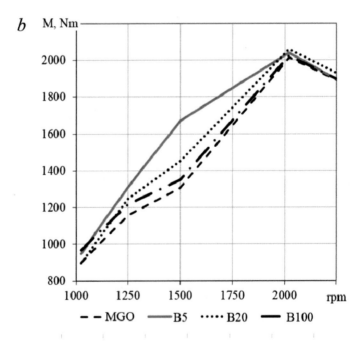

Fig. 2.9 Dependence of the characteristics of the CAT 3176C engine operating on DF, BD, and their mixtures on the rotation frequency: **a**—power; **b**—rotation torque

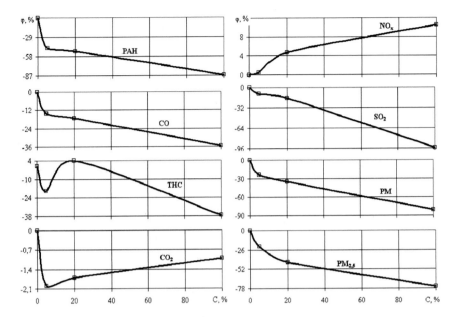

Fig. 2.10 Dependence of the engine CAT3176C emissions level on the biodiesel fuel concentration in the mixture (Emissions at operation on DF Are taken as 0)

by 5–7%, as well as decrease in efficiency and specific fuel consumption [38, 48, 93–97].

During the CAT 3176C marine engine testing (Fig. 2.9b), the rotation torque at all rotational speeds was lower for the DF than that for the B100 and its mixtures. On average, the value of the torque increases by 7.5% for B5, by 4.2% for B20, and by 2.5% for B100 when compared to the engine operation on diesel fuel [63].

Figures 2.10 and 2.11 depict the dependence of the change in the emissions level for the marine engine CAT 3176C when operating on B5, B20, B100. The data are presented in a relative form [63].

Nitrogen oxides are the only emission components, the level of which increases with the BD use in a practically linear dependence on the biodiesel fuel concentration: 0.5% for B5, 4.7% for B20 and 10.5% for B100. The use of a special calculation model in the work showed that, the level of CO_2 emissions is reduced by 3.5 kg/l of fuel when using B100, by 0.73 kg/l for B20, and by 0.31 kg/l for B5 [63]. The calculation was based on the analysis of the life cycle of fuel (from growing the raw materials for the BD to its combustion in the engine). Taking into account the emissions level only during engine operation, CO_2 is reduced by 1% for B100, 1.7% for B20 and 2% for B5 compared to DF; emissions reduction is observed in all operating modes. RM emissions are reduced by 23%, 35% and almost 82% for B5, B20 and B100, respectively. The level of CO emissions is reduced by 14%, 17% and 35% for B5, B20 and B100, respectively, if compared to the DF. The decrease in CO is not proportional to the BD concentration in the mixture and has a nonlinear

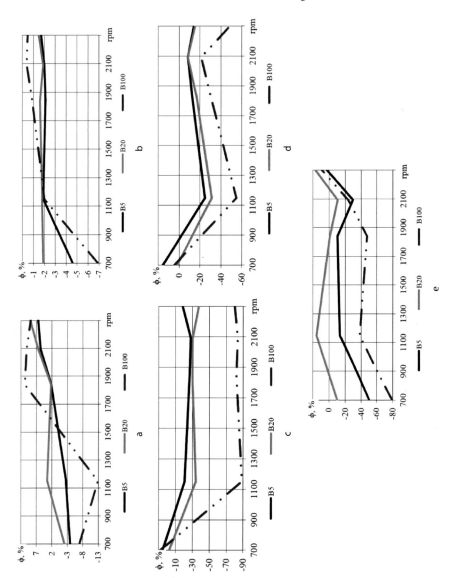

Fig. 2.11 Impact of the rotation frequency on the emissions concentration at the engine CAT 3176C operation on BD and its mixtures (Emissions for that on DF are taken as 0): **a**—NOx; **b**—CO_2; **c**—PM; **d**—CO; **e**—THC

dependence. A significant decrease in carbon monoxide is observed even with the small content of biodiesel (5%). The THC emissions decrease on average by 19 and 37% for B5 and B100, but increase by 4% for B20. Emissions of hydrocarbons are reduced in almost all modes of engine operation, except for the full throttle mode [63].

Table 2.9 Change in the level of EG pollutants emissions during biodiesel fuel and B20 mixture combustion

Pollutants	B100 (%)	B20 (%)
Hydrocarbons	−67	−20
CO	−48	−12
Solid particles	−47	−12
NO_x	+10	+2
Sulphur	−	−20

Comparison of the environmental characteristics (averaged values) of the engine working on biodiesel fuel by contrast with the diesel fuel is given in Table 2.9 [10].

Figure 2.12 shows the results of testing of the two-cylinder engine RD270 Ruggerini running on the BD mixtures produced from frying oil. The engine operation on the BD and DF mixtures practically coincides in character with work on pure diesel fuel. Compared to the use of pure diesel fuel, the average power value increases at the engine running on the B20 (the most) and B40 mixtures (Fig. 2.12a). The difference between the power values, both upward and downward, is relatively small (± 3%) [96].

The nature of the relative change in the specific fuel consumption when the RD270 Ruggerini engine operates on the DF and BD mixtures is shown in Fig. 2.12b. Depending on the rotation frequency, the specific fuel consumption for BD and DF mixtures can both increase and decrease, but on average it is higher (by 0.6–4%) for the mixtures compared to the operation on diesel fuel. If using the B40 g_e, it averagely decreases by 1.4%. Fuel consumption decreases the most at the engine operation on B10—this indicator is up to 11.5% at maximum speed. The g_e decreases for all mixtures compared to diesel fuel at $n = 2000$ min^{-1} and $n = 2800$ min^{-1} [96].

The average value of rotation torque decreases while using the B10 and B50 (by 1% and 0.8%, respectively). For all other mixtures compared to DF, it increases up to 2.5% (for B20 the most) for the RD270 Ruggerini engine (Fig. 2.12c). The nature of the change in the rotation torque is the same for both DF and for all the mixtures considered. The characteristic was determined for all fuels under full engine load [96].

Table 2.10 and Fig. 2.13 provide the comparison of the characteristics of the Kubota V1305 diesel engine when operating on DF, B100 and B20.

During the Kubota V1305 engine testing, the average power increased by 2% for the B20 and decreased by 7% for the B100 (again, compared to the MGO). The increase in g_e by 7% and 20% is fixed for B20 and B100, respectively. At the same time, the engine efficiency is 29.6% for MGO, 28.3% for B20 and 28.2% for B100 [95].

When the engine is running on biodiesel, power is reduced due to the lower calorific value. Continuous operation of the engine on this fuel leads to the formation of lacquer deposits on the fuel injectors, corrosion and wedging of the internal components of the fuel injection system, malfunction of the fuel pump due to water

Fig. 2.12 Dependence of
the RD270 Ruggerini engine
characteristics on the
rotation frequency during its
operation on DF (MGO), BD
and their mixtures:
a—power; **b**—specific fuel
consumption; **c**—rotation
torque

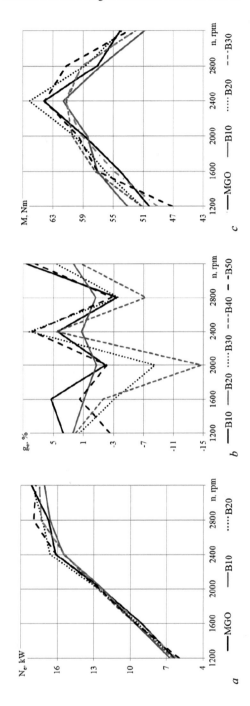

Table 2.10 Characteristics of the Kubota V1305 engine operating on MGO, B20 and B100

Fuel	Fuel consumption, kg/h	Average power, kW	Specific fuel consumption, g/(kW · h)	Efficiency %
MGO	5.92	22,5	263	29.6
B20	6.46	23	281	28.3
B100	6.92	21	330	28.8

ingress, formation of sludge and sediment, which leads to the decrease in engine durability [46].

Despite some differences, the general trends in the change of the characteristics of the working process and environmental performance of the engines during their transition from petroleum fuels to WFE, LNG, BD and its mixtures largely coincide.

2.5 Conclusions

1. The potentially productive alternative fuels that have become widespread nowadays include the following: alcohol fuels (mainly methanol and ethanol are used instead of gasoline or mixed with it); ethers (dimethyl and diethyl as additives to gasoline), hydrogen (as an additive to traditional fuels, independent energy carrier for engines of a special design, for use in the fuel cells); synthetic liquid fuels from hydrocarbon raw materials; coal-origin fuels (solid, liquid, gaseous); composite fuels (the most promising ones from the environmental point of view, they are the derivatives of coal or oil and plastic waste that are difficult to dispose); biofuel (solid, liquid, gaseous); colloidal fuels (water-coal and water-fuel emulsions based on petroleum fuels).
2. When it comes to fuels used onboard of the ships, there is experience in using the following ones: the liquefied petroleum gas; the fuel cells (most often methanol is used as a source of hydrogen, while diesel fuel and natural gas are used less often); water-fuel emulsions (mainly water–oil and water-diesel); natural gas (can be stored onboard of the vessel in compressed and liquefied states); biodiesel fuels.
3. Presently, the most promising and real alternatives to traditional marine fuels are biodiesel fuels and their mixtures, water fuel emulsions, and liquefied natural gas. There are several reasons for this. These fuels are produced or extracted on the industrially significant scale and adapted for combustion within marine power equipment. They are non-toxic, and the environmental efficiency improves as a result of their use. Their price is slightly higher and sometimes lower than the cost of traditional fuels, and there is a positive experience in using these fuels onboard of the ships.

Fig. 2.13 Change in the
characteristics of the Kubota
V1305 engine operating on
MGO, B20 and B100:
a—rotation torque;
b—effective power

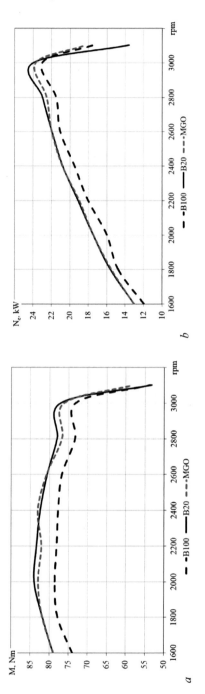

4. The analysis of the experience of using the fuels listed above onboard showed that the following changes are required for the effective SPP operation:
   – insertion of the additional equipment in the fuel system for the preparation of emulsified fuel by cyclic or continuous methods;
   – availability of the separate fuel system for LNG, which must include cryogenic pipelines, liquefied gas storage tanks, and regasification equipment; the tanks must be located not in the engine room, but in a separate room;
   – for the effective use of biodiesel fuels, it is recommended to change the parameters of the main equipment of diesel fuel systems and the configuration of individual sections, while maintaining the standard configuration of the system and taking into account the difference in physical and chemical and rheological characteristics. It is necessary to use structural and sealing materials that are highly compatible with the BD.
5. The analysis of the results of experimental studies indicates that the working process of internal combustion engines using WFE, LNG and BD has a number of features and can vary depending on the characteristics of the feedstock for fuel (primarily biodiesel fuels and their mixtures), as well as engine modifications (gas-diesel, dual-fuel, gas-piston, MSE, HSE). The application of these fuels makes it possible to improve the environmental characteristics of the engines and reduce the level of all the main pollutants contained in the exhaust gases. The CO and hydrocarbon emissions can be increased at the LNG use, and some increase in nitrogen oxide emissions is observed for the BD fuel. There may be a reduction in engine power due to the calorific value of WFE and BD being lower than that of petroleum fuels; the calorific value of LNG is higher, but due to the low density, the amount of energy introduced with a unit of fuel volume is less, which can also lead to power reduction. It is possible to eliminate this drawback by making necessary changes in the design of diesel engines or reconfiguring the fuel equipment.

References

1. N. V. Golubev, Proektirovanie energeticheskih ustanovok morskih sudov [Designing of power plants for marine vessels]. Leningrad: Sudostroenie Publ., 1980, 312 p.
2. G. A. Artemov, V. P. Voloshin, A. Ya. Shkvar, V. P. Shostak, Sistemyi sudovyih energeticheskih ustanovok [Ship power plant systems]. Leningrad: Sudostroenie Publ., 1990, 376 p.
3. G. Benvenuto, M. Figari, C. M. Rizzo, "Effects of fuel quality on two strokes slow speed engines", in Maritime transportation and exploitation of ocean and coastal resources: 11th International Congress of the International Maritime Association of the Mediterranean, Lisbon, 2006, pp. 627–634.
4. B. M. Gorbov, Enerhetychni palyva [Power-generating fuel]. Mykolayiv: UDMTU Publ., 2003, 328 p.
5. V. V. Voznitskiy, Praktika ispolzovaniya morskih topliv na sudah [Practice of marine fuels using on ships]. Sankt-Peterburg: Biblioteka sudovogo mehanika Publ., 2006, 124 p.
6. B. N. Kuznetsov, "Motornyie topliva iz alternativnogo nefti syirya [Motor fuels from materials alternative to fossil oil]", Sorosovskiy obrazovatelnyiy zhurnal, vol. 6, pp. 51–56, # 4, 2000.

7. Cleaner fuels: Policy guidelines for vehicle emission in Asia (2003) [Online]. Available: https://www.adb.org/sites/default/files/publication/171733/cleaner-fuels.pdf
8. Ya. B. Chertkov, Perspektivnyie avtomobilnyie topliva [Perspective automobile fuels]. Moskva: Transport Publ., 1982, 319 p.
9. Ya. B. Chertkov, Motornyie topliva [Motor fuels]. Novosibirsk: Nauka Publ., 1987, 207 p.
10. B. P. Pundir, Engine emissions: pollutant formation and advances in control technology. Oxford: Alpha Science International Ltd, 2007, 301 p.
11. A. M. Danilov, Primenenie prisadok v toplivah dlya avtomobiley [Usage of fuel additives for vehicles]. Moskva.: Himiya Publ., 2000, 232 p.
12. "Volvo Multi Fuel", Brennstoffspiegel und Mineralölrdsch, p. 6, #8, 2006.
13. S. Hauri, "Vorteil für Wankelmotor", Automobile Revue, p. 11, #35, 2006.
14. Modern shale gas development in the United States: a primer (2009) [Online]. Available: https://energy.gov/sites/prod/files/2013/03/f0/ShaleGasPrimer_Online_4-2009.pdf.
15. M. Ridely (2011). The shale gas shock [Online]. Available: https://www.marcellus.psu.edu/resources/PDFs/shalegas_GWPF.pdf.
16. A. M. Osipov, A. F. Popov, S. V. Grischuk and others, "Proizvodstvo sinteticheskogo topliva sovmestnoy pererabotkoy iskopaemyih ugley i othodov plastmass (obzor) [Alternative fuels manufacturing by utilizing co-processing of fossil coals and plastics scrap (review)]", Ekotehnologii i resursosberezhenie, pp. 26–34, #4, 2004.
17. B. N. Kuznetsov, "Novyie podhodyi v himicheskoy pererabotke iskopaemyih ugley [New approaches in chemical treatment of fossil coals]", Sorosovskiy obrazovatelnyiy zhurnal, pp. 50–57, #6, 1996.
18. "Coal to liquids: the time is right", Electric perspectives, vol. 32, pp. 74–75, #1, 2007.
19. T. A. Zheleznaya, G. G. Geletuha, "Sovremennyie tehnologii polucheniya zhidkogo topliva iz biomassyi byistryim pirolizom [Modern technologies of manufacturing of liquid fuel from biomass made by fast pyrolysis]", Promyishlennaya teplotehnika, pp. 91–100, #4, 2005.
20. S. S. Rizhkov, M.V. Rudyuk, "InnovatsIynI tehnologiyi utilizatsiyi organichnih vidhodiv z otrimannyam alternativnogo paliva na osnovi bagatokonturnogo tsirkulyatsiynogo pirolizu [Innovation technologies of organic waste recycling with alternative fuel manufacturing based on multistage circular pyrolysis]", Zbirnik naukovih prats NUK, pp. 133–142, #2, 2010.
21. A. Sestan, Z. Parat, "Development possibilities of modern coal-fired marine propulsion plant", in Maritime transportation and exploitation of ocean and coastal resources: 11th International Congress of the International Maritime Association of the Mediterranean, Lisbon, 2006, pp. 699–704.
22. V. M. Gorbov, Entsyklopediya sudnovoyi enerhetyky [Encyclopedia of marine engineering]. Mykolayiv: NUK Publ., 2010, 624 p.
23. M. D. Mamedova, Rabota dizelya na szhizhennom gaze [Operation of diesel engine on liquefied gas]. Moskva: Mashinostroenie Publ., 1980, 149 p.
24. "New study solves environmental concerns", Ship and Boat International, pp. 11, March/Apr, 2007.
25. "Water taxis to run on hydrogen gas", Ship and Boat International, pp. 50, May/June, 2007.
26. N. Hallale, I. Moore, D. Vauk, "Hydrogen: liability or asset?", CEP, pp. 66–75, #9, 2002.
27. "Fuel cells take small steps toward reality", The Naval Architect, pp. 25–26, #3, 2007.
28. V. M. Gorbov, Akusticheskaya obrabotka topliva dlya sudovogo energeticheskogo oborudovaniya [Acoustic treatment of fuel for ship power equipment], Nikolayiv: NKI Publ., 1989, 51 p.
29. V. M. Gorbov, Primenenie vodotoplivnyih emulsiy v sudovoy energetike [Water fuel emulsion usage in marine engineering]. Nikolayiv: NKI Publ., 1991, 54 p.
30. S. P. Zubrilov, V. M. Seliverstov, M. I. Braslavskiy, Ultrazvukovaya kavitatsionnaya obrabotka topliv na sudah [Ultrasonic cavitation fuel treatment on ships]. Leningrad: Sudostroenie Publ., 1998, 80 p.
31. O. N. Lebedev, V. A. Somov, V. D. Sisin, Vodotoplivnyie emulsii v sudovyih dizelyah [Water fuel emulsion in diesel engines]. Leningrad: Sudostroenie Publ., 1988, 108 p.
32. "Cleaner four strokes", Propulsion, pp. 42–44, 2007.

33. V. A. Zagoruchenko, A. M. Zhuravlev, Teplofizicheskie svoystva gazoobraznogo i zhidkogo metana [Thermophysical properties of gas and liquid metan]. Moskva: Izdatelstvo standartov Publ., 1969, 236 p.
34. "Bergen hits target", Marine Power and Propulsion (a one–year subscription to The Naval Architect), p. 30, 2007.
35. "Ro-Ro for the future", Propulsion, pp. 10–12, 2008.
36. O. Ashpina. "Raps—kultura strategicheskaya [The rape is a strategic cropper]", The Chemical Journal, pp. 40–44, #9, 2005.
37. S. N. Devyanin, V. A. Markov, V. G. Semenov, Rastitelnyie masla i topliva na ih osnove dlya dizelnyih dvigateley [Seed oils and fuels based on it for diesel engines]. Harkiv: Novoe slovo Publ., 2007, 452 p.
38. V. Smaylis, V. Senchila, K. Bereyshene, "Motornyie ispyitaniya RME na vyisokooborotnom dizele vozdushnogo ohlazhdeniya [Engine tests of RME for highspeed engine with air cooling]", Dvigatelstroenie, pp. 45–49, #4, 2005.
39. R. Berryman (2008) Biodiesel in marine applications [Online]. Available: https//www.seismi cevents.ca/biodieselsymposium2008/pdf/R_Berryman.pdf.
40. D. L. Clements (1996) Blending rules for formulating biodiesel fuel [Online]. Available: https://biodiesel.org/reports/19960101_gen-277.pdf.
41. Biodiesel handling and use guidelines (2009) [Online]. Available: https://biodiesel.org/docs/using-hotline/nrel-handling-and-use.pdf?sfvrsn=4.
42. Biodiesel cold weather blending study (2005) [Online]. Available: https://biodiesel.org/docs/default-source/ffs-performace_usage/cold-weather-blending-study.pdf?sfvrsn=6.
43. J. Van Gerpen (1996) Cetane number testing of biodiesel [Online]. Available: https://images.rcuniverse.com/forum/upfiles/45069/Id95843.pdf.
44. S. Lebedevas, A. Vaicekauskas, "Research into the application of biodiesel in the transport sector of Lithuania", TRANSPORT, vol. XXI, pp. 80–87, #2, 2006.
45. M. M. Conseicao, R. A. Candeia, H. J. Dantas and others, "Rheological behavior of castor oil biodiesel", Energy & Fuels, pp. 2185–2188, #19, 2005.
46. "Coping with biodiesel—the new challenge", MER, pp. 18–20, Sept. 2007.
47. B. Holden (2006) Effect of biodiesel on diesel engine nitrogen oxide and other regulated emissions [Online]. Available: https://www.jgsee.kmutt.ac.th/jsee/JSEE%202012/PDF%20f ile%20JSEE%203(1)%202012/11.Effect%20of%20biodiesel%20pp.%2035-47.pdf.
48. Y. X. Li, N. B McLaughlin., B. S. Patterson, S. D. Burtt, "Fuel efficiency and exhaust emissions for biodiesel blends in an agricultural tractor", Canadian biosystems engineering, vol. 48, pp. 15–22, 2006.
49. Lubricity benefits (1998) [Online]. Available: https://biodiesel.org/docs/ffs-performace_usage/lubricity-benefits.pdf?sfvrsn=4.
50. R. L. McCormick, T. L. M. Alleman, Ratcliff and others (2005) Survey of the quality and stability of biodiesel and biodiesel blends in the united states in 2004 [Online]. Available: https://biodiesel.org/reports/20051001_gen356.pdf.
51. The physical & chemical characterization of biodiesel low sulfur diesel fuel blends (1995) [Online]. Available: https://biodiesel.org/reports/19951230_gen-253.pdf.
52. R. von Wedel (1999) Technical handbook for marine biodiesel in recreational boats [Online]. Available: https://biodiesel.org/reports/19990401_mar-015.pdf.
53. T. A. Zhelyezna, "Stan rozvytku ta perspektyvy vyrobnytstva i zastosuvannya ridkykh palyv z biomasy. Ch. 2 [State-of-the-art and manufacturing and usage perspectives for liquid fuels made from biomass. Part 2]', Эkotekhnolohyy y resursosberezhenye, pp. 3–8, #2, 2004.
54. "A cheaper way to lay the foundations", Offshore Marine Technology, pp. 12–16, 4th Quarter, 2011.
55. "A little ray of sunshine", The Naval Architect, pp. 35, Jan. 2009.
56. "Green cruises ride winds of change", The Naval Architect, pp. 12, Oct. 2009.
57. M. Berisa, "New cruise ship concept from Wartsila (Part 1)", Twentyfour7, pp. 51–54, #2, 2007.
58. "Fuel cell boat moves towards reality", The Naval Architect, pp. 26, May 2009.

59. "Fuel cell ship in the real world", The Naval Architect, pp. 56–57, Nov. 2008.
60. M.-H. Bård (2002) Fuel cell technology for ferries [Online]. Available: https://www.sintef.no/globalassets/upload/marintek/pdf-filer/publications/fuel-cell-technology-for-ferries_bmh.pdf.
61. "Viking Lady tests fuel cell power", Marine Power & Propulsion (a one-year subscription to The Naval Architect), pp. 30, 2009.
62. "A biological CAT", MER, p. 24, Sept. 2007.
63. BioMer: biodiesel demonstration and assessment for tour boats in the Old Port of Montreal and Lacine canal national historic site (2005) [Online]. Available: https://www.sinenomine.ca/Download/BioMer_ang.pdf.
64. P. Einang, "A gas evolution", Propulsion, pp. 50–56, 2007.
65. D. Zbaraza (2004) Natural gas use for on-sea transport [Online]. Available: https://www.ipt.ntnu.no/~jsg/studenter/diplom/DawidZbaraza2004.pdf.
66. V. M. Gorbov, V. S. Mitenkova, "Osobennosti toplivnyih sistem SEU na prirodnom gaze [Characteristic properties of ship power plants fuels systems using liquefied natural gas]", Aviatsionno-kosmicheskaya tehnika i tehnologiya, pp. 20–24, #7, 2008.
67. M. Berisa, "New cruise ship concept from Wartsila (Part 2)", Twentyfour7, pp. 51–54, #3, 2007.
68. "Clean Cruising Concept", Diesel & Gas Turbine Worldwide, pp. 59–61, Apr. 2008.
69. P. Einang (2007) LNG som drivstoff for skip [Online]. Available: https://www.gasforeningen.se/upload/files/seminarier/gasdagarna2007/foredrag/per%20magne%20eingang.pdf.
70. "Gas breakthrough in merchant sector", Rolls–Royce merchant supplement (a one-year subscription to The Naval Architect), pp. 3, 2008.
71. "Hybrid coast guard vessel can also burn LNG", Warship technology, pp. 18–21, Oct. 2009.
72. O. Levander, "Cruising on gas into a cleaner future", Wärtsilä technical journal, pp. 26–31, #1, 2007.
73. "LNG–fuelled cargo ship design contract", MER, pp. 37, Nov. 2008.
74. "LNG–fuelled vessels increase in popularity", Ship and Boat International, p. 22, May/June 2007.
75. "New gas ferry for Norway", The Naval Architect, pp. 58, Apr. 2009.
76. T. G. Osberg (2008) Gas Fuelled Engine Installations in Ferries. Emission Reductions and Safety Considerations [Online]. Available: https://www.sname.org/HigherLogic/System/DownloadDocumentFile.ashx?DocumentFileKey=be40b7e0-654f-44f4-a3d1-b1a8fa293c9b.
77. J. Paananen, LNG fuelled ships and auxiliaries. Rostock: Wärtsilä ship power Publ., 2007, 44 p.
78. J. Pagni, "New, greener LNG tug concept", Twentyfour7, pp. 64–65, #3, 2008.
79. "Shortsea ships use gas to go green", The Naval Architect, pp.63, Jan. 2009.
80. A. Skjervheim (2008) The LNG in ship experience—Can we bring in biogass? [Online]. Available: https://www.sgc.se/nordicbiogas/resources/Aksel_Skjervheim.pdf.
81. D. Tinsley "Clean, green and lean", Shipping World & Shipbuilder, pp. 20–27, Nov. 2008.
82. D. Tinsley "The gateway to an emergent power", Shipping World & Shipbuilder, pp. 21–24, Oct. 2008.
83. "The answer to marine emissions", Propulsion, pp. 10–12, 2010.
84. Wärtsilä engine products (2010) [Online]. Available: https://ip2010.turkuamk.fi/pdf/Wartsila_engine_products.pdf.
85. "Wartsila 50DF—performance optimized for dual–fuel operation", Energy news, pp. 10–13, #9, 1999.
86. "A very green Emerlad", The Naval Architect, pp. 8–10, July/Aug. 2007.
87. "Dual fuel engines make mark on LNG", The Naval Architect, pp. 15, March 2007.
88. "Groundbreaking 51/60DF announcement", DIESELFACTS, pp. 1, #2, 2007.
89. "International cooperation on ship design. MAN engines breakthrough", The Naval Architect, pp. 15, June 2007.
90. "LNG power, your flexible friend", The Naval Architect, pp. 54–57, Feb. 2010.
91. "MAN breaks through with dual fuel option", The Naval Architect, pp. 47, #3, 2008.
92. "Wartsila doubles its four–stroke", The Naval Architect, pp. 77, Feb. 2008.

93. I. P. Vasilev, "Rezultatyi ispyitaniy v dizelnom dvigatele smesey topliv rastitelnogo proishozh-deniya [Results of diesel engine tests operating on biofuel blends]", Ekotehnologii i resursos-berezhenie, pp. 3–9, #2, 2007.

94. D. Bibic, A. Hribernik, I. Filipovic, B. Kegl, "Influence of alternative fuels on combustion indicators with diesel engines", Goriva i maziva, vol. 46, pp. 205–222, #3, 2007.

95. Biodiesel engine testing (2005) [Online]. Available: https://circle.ubc.ca/bitstream/handle/2429/22426/Biodiesel%20Engine%20Testing.pdf?sequence=1.

96. G. Najafi, B. Ghobadian, T. Yusaf, H. Rahimi, "Combustion analysis of a CI engine performance using waste cooking biodiesel fuel with an artificial neural network aid", American Journal of Applied Sciences, pp. 756–764, #4 (10), 2007.

97. T. Raj (2008) Comparison of operational characteristics of diesel engine run by bio diesel (rubber seed oil) with diesel fuel operation [Online]. Available: https://www.ese.iitb.ac.in/~ica er2007/Latest%20PPT%20File/48_023_TS4%20A.pdf.

Chapter 3
Liquefied Natural Gas as Marine Fuel

3.1 Review of the Fuel Systems for LNG-Fueled Power Plants

Liquefied natural gas as fuel is used on ships and heavy vehicles, as well as stationary heat supply facilities.

The basic equipment of the fuel system of power and transportation facilities includes: a receiving cryogenic pipeline, a cryogenic fuel supply pump, LNG storage tanks, regasifiers (one or more heat exchangers or electric heaters for gas evaporation and heating), other equipment (filters, fuel distribution devices, buffer tanks, etc.) [1, 2]. The configuration of fuel systems can vary depending on the field of application.

When using natural gas as the fuel onboard of the ship, the structure and equipment of the SPP fuel system is to be radically changed. It should be considered that it takes two times more volume to store LNG than to store marine diesel oil (MDO) onboard of the vessel with the same energy efficiency of the main engines, and storing liquefied or compressed natural gas (CNG) requires 5 times more volume.

The mass of fuel tanks with fuel is by 1.5 times higher for LNG than for MDO, and by 4 times for compressed natural gas (Fig. 3.1) [3].

The main methods of the LNG storage include: isothermal—storage at a constant temperature, providing the overpressure of saturated vapor close to atmospheric; semi-isothermal—storage of gas in tanks at a constant temperature and saturated vapor pressure above atmospheric. The maximum technologically justified pressure of the semi-isothermal storage method is 1.2 MPa (in some cases 1.6 MPa for stationary facilities) for LNG complexes [4].

The onboard liquefied gas storage systems of the ships and their configuration depend on the specific operating conditions, duration of the voyage and a number of other factors. The analysis of the circuits and composition of the LNG fuel system equipment is of interest. To date, a number of circuit solutions of the liquefied gas fuel systems for vehicles and stationary storage facilities have been recorded. Figures 3.2, 3.3, 3.4, 3.5, 3.6 and 3.7 present the diagrams of LNG fuel systems of various units.

© Shanghai Scientific and Technical Publishers 2021
X. Yang et al., *Alternative Fuels in Ship Power Plants*,
https://doi.org/10.1007/978-981-33-4850-9_3

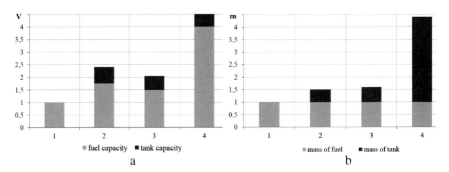

Fig. 3.1 Relative parameters of the fuel systems when using MDO (1), LNG under pressure of 0.2 MPa (2), LNG under pressure of 1.0 MPa (3) and CNG under pressure of 20.0 MPa (4): **a** Volume of fuel and storage tanks; **b** Mass of fuel and storage tanks

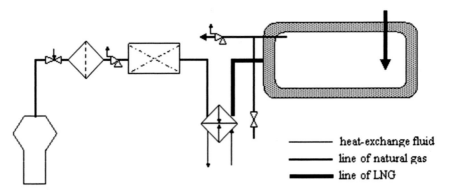

Fig. 3.2 Structural diagram of the fuel system at the engine operation on LNG: 1—engine; 2—gas regulator, 3—fuel filter; 4—shut-off valve; 5—buffer tank; 6—evaporator; 7—relief valve (gas relief); 8—regulating valve; 9—LNG storage tank; 10—check valve

Figure 3.2 shows the diagram of the fuel system, which can be used for both compressed and liquefied gas [1]. This scheme is quite universal. It can be used as a basic scheme for the fuel systems of powers plants of bi-fuel engines [1].

The diagram of the LNG evaporation port terminal with the subsequent supply of natural gas to the main gas pipeline provides the delivery of liquefied gas by gas carriers and storage in stationary large-scale land tanks (Fig. 3.3). A portion of the gas evaporating naturally during heating of the storage tanks may be supplied by the compressor to the main gas pipeline or sent to the gas turbine for disposal. Forced evaporation of LNG is provided by feeding it to evaporators, and then through the gas pipeline to consumers. Seawater is used as the coolant; it is supplied by the pumps, which can be used to cool the condensers of steam turbines in the system of deep heat recovery if the provision is made with the diagram. The naturally evaporated gas must be continuously discharged from the tank to maintain the pressure inside the tank at the constant level [5].

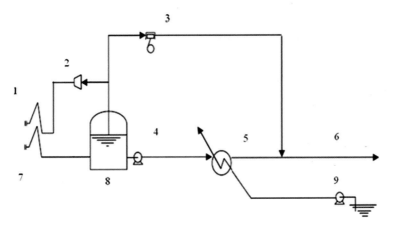

Fig. 3.3 Diagram of the port LNG regasification unit: 1—turbine exhaust gases; 2—utilization gas turbine; 3—natural gas vapor supply compressor; 4—LNG pump; 5—LNG evaporator; 6—gas supply to the main gas pipeline; 7—discharge of the liquefied phase; 8—LNG storage tank; 9—coolant (seawater) supply pump

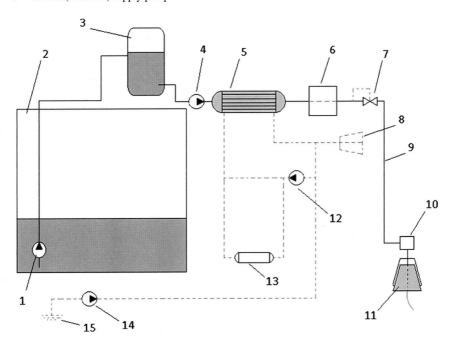

Fig. 3.4 Diagram of the gas carrier regasification unit: 1—cryogenic pump; 2—LNG storage tank; 3—reduced-pressure container; 4—high-pressure cryogenic pump; 5—LNG evaporator; 6—batcher; 7—regulating valve; 8—seawater discharge; 9—high-pressure hose; 10—articulated joint; 11—submerged turret buoy; 12—circulating pump; 13—steam heater; 14—ballast pump; 15—seawater intake

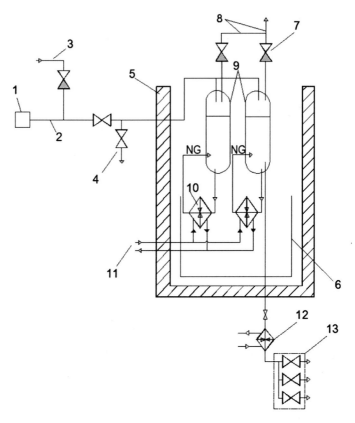

Fig. 3.5 Schematic diagram of the SPP fuel system when using LNG and the tanks placement within the vessel superstructure: 1—bunkering device; 2—cryogenic receiving pipeline; 3—purging with nitrogen; 4—LNG withdrawal to another group of tanks; 5—fire-proof tank protection casing; 6—drip tray; 7—gas discharge valve; 8—evaporation gas discharge line; 9—cryogenic LNG storage tanks; 10—gas evaporator for maintaining the pressure in the tanks; 11—coolant drainage and supply to the evaporator; 12—LNG evaporator-heater; 13—gaseous fuel distribution unit

In most cases, LNG gasification occurs within the port facilities, but since 2005 several gas carriers have been built with similar facilities onboard (Fig. 3.4) [6].

The LNG transfer from the liquid phase to the gaseous phase and its subsequent supply to the shore through the main gas pipeline takes place directly onboard of the vessel. A shell-and-tube heat exchanger is used as the evaporator. Seawater serves as the coolant, which is supplied by the ballast pump located in the engine room. Seawater after the evaporator is discharged overboard. At low temperatures of seawater, heating of the coolant is provided before its entering the evaporator to prevent its freezing there. As a heating medium, the heater uses steam from auxiliary or main boilers (for steam turbine power plants) of the vessel. In addition, it is provided that the fresh water heated in the steam heater and circulating in the closed circuit can be fed into the LNG evaporator [6].

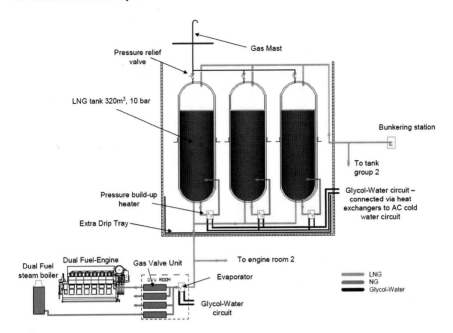

Fig. 3.6 Liquefied natural gas system on a cruise ship

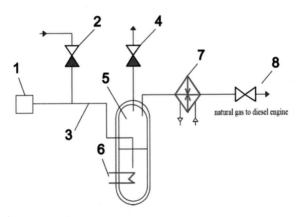

Fig. 3.7 Schematic diagram of the SPP fuel system for diesel generators when using LNG and vertical placement of tanks: 1—bunkering device; 2—purging with nitrogen; 3—storage tank for liquefied gas; 4—discharge of vaporized gas from the tank; 5—gas heater; 6—fuel to consumers; 7—receiving pipeline; 8—immersion evaporator of the fuel liquid phase

The company Wärtsilä developed an alternative version with the standard arrangement of the LNG tanks in the vessel hull. This development involves the use of tanks with low storage pressure, made of B-type stainless steel according to the International Code for LNG Tanks. Such containers are cheaper and take less space, since they have the shape of a parallelepiped. Yet, according to existing classification requirements, the use of such tanks is allowed only on LNG carriers [7–9]. LNG from the tanks is supplied to the heat exchanger, where the liquid evaporates. The gas goes to the engines through the control valves. The liquid phase enters the evaporators under gravity due to the difference in levels (the tanks are located above the heat exchangers) and the excess pressure in the tanks, which ensures the simplicity and reliability of the system. The pressure in the tank is maintained by evaporating a certain amount of LNG in the separate evaporator for each tank, located in the same room as the tanks.

Two evaporators and distribution blocks are located in separate rooms next to the heat exchangers. Six evaporators are provided for six tanks with LNG, each in the separate room. A "water–glycol" mixture is used to heat LNG within the heat exchangers. To increase energy efficiency, this cooled mixture is used in the air conditioning system circulation circuit. In such a way, low-potential energy is utilized (Fig. 3.5) [9–11].

The diagram is used when the fuel tanks are placed within the ship superstructure. Containers with gas are located vertically, although it is also possible to arrange them horizontally (such as on container ships), necessarily in a protective casing to ensure fire safety. When placed in the superstructure, 6–8 tanks of a relatively small volume are used; they are usually located in two separate rooms of 3–4 tanks each. Under the containers, there is a metal drip tray. Individual evaporators are provided for each tank to maintain the required pressure because it decreases during the process of fuel sampling. According to the diagram, one LNG regasification unit is provided per group of tanks; it is not advisable to install evaporators-heaters for each container because of their relatively small volume. The regasification unit is located in the hull of the vessel in the separate room near the engine room. LNG is delivered to the evaporators under gravity through reliably insulated cryogenic pipelines. There is also a distribution valve block for the natural gas supply to different consumers, engines, boilers (if the bi-fuel boiler is installed).

The vessel "Viking Energy" for the servicing of gas-producing platforms is equipped with bi-fuel engines powered by natural gas and diesel fuel. LNG is stored in reliably protected large-size tanks in the central part of the vessel. The tank is a horizontal cylinder with rounded ends, made of high-quality stainless steel. The tank consists of two chambers, internal and external; the distance between them is 300 mm. There is a reliable deep-vacuum insulation between them. LNG is stored at −162 °C.

The company Wärtsilä has developed an LNG fuel system for cruise ships (Fig. 3.6). Liquefied natural gas is stored on board in 6 cryogenic tanks with the capacity of 320 m^3 at the pressure of 1 MPa. The tanks are provided with exhaust valves, through which the gaseous phase is released in case of excessive pressure increase. Each fuel tank is equipped with heat exchangers for pressure maintaining. LNG is regasified in evaporators (2 evaporators for 6 tanks) before its supply to the consumer. The "water–glycol" mixture is used as the coolant for pressure maintaining in the evaporators and heat exchangers; after cooling, it is fed to the air-conditioning system [10].

The volume of the reservoir is 234 m³; the useful volume is 220 m³ when filled with the fuel up to the acceptable level. The tanks are under pressure. There is a built-in evaporator of the coil type. The tanks are installed in fire protection room of the A60 class. All gas pipelines and valves are placed in ventilated casings with emergency sensors signaling the leak; places where gas can accumulate are under constant monitoring [3].

On the ferry "Glutra", LNG is stored in two steel cryogenic tanks with the maximum capacity of 27.2 m³ with vacuum pearlite insulation. The filling factor of each tank is 0.85. Tanks are located under the upper deck in the sheet steel boxes with double walls and special filling to reduce vibration.

Refueling occurs every 5–6 days for 2 h. LNG is delivered to the port in the cryogenic tank mounted on a truck; filling takes place via a receiving connection device on the ferry. Liquefied gas is supplied from the cargo tank with steam under pressure created by the air evaporator installed on the truck. The LNG can be poured into the bottom or injected into the top of the fuel tanks. Typically, injection is used to create a gas "cushion" above the liquid phase so that there is no overpressure, which is created in the tank during the LNG filling process.

Before use, the liquefied gas must be regasified and sent to the engines at the temperature about 20 °C under the pressure of 0.5 MPa. The regasification process takes place in the evaporator of hot water installed on the gas tank. The hot water of the ship systems goes through two coils: one unit with the capacity of 390 kW evaporates 600 m³ of gas per hour; the second unit is less powerful. It evaporates the LNG in the amount sufficient to maintain pressure in the tank. The connection for bunkering is located on the starboard side of the vessel. For safety purposes, it is filled with inert gas (nitrogen). Under regular conditions, the vessel's bunkering with gas takes three hours once a week (bunkering without preparatory work takes about 2 h) [3].

The presence of a submerged evaporator located at the tank bottom is a special feature of the diagram in Fig. 3.7. The coil evaporator can have the form of an electric or heat exchanger using saturated steam as the coolant. The submerged heater simultaneously serves to maintain the pressure in the tank by evaporation of the small amount of liquid and as the first stage of the regasification unit, evaporating part of the liquid. After that, the saturated steam is supplied to the external heater where the gas is heated to the required temperature for its supply into the engine [12].

All of the mentioned above makes it possible to formulate the basic requirements for the development of the SPP fuel system diagrams for LNG.

The schematic diagrams of such systems differ in location, quantity and position of containers (vertical or horizontal), number of heat exchangers, and organization of the regasification process. Tanks of a small volume may be located in the vessel superstructure or on the deck. The choice of their horizontal or vertical placement will depend on the type and structure of the vessel, the availability of free space, and safety requirements. With this arrangement, the number of tanks can be 2...8 pieces; they are placed in special protective casings. The bunkering pipeline is located below the filling line of the tanks. As a result, the hydraulic resistance and pump head increase.

If the number of tanks is more than four, it is advisable to install one regasification unit for 2…4 tanks. In general, there should be at least two evaporators on the vessel.

Submerged evaporators are convenient to use in vertical tanks with a relatively small volume of fuel contained in them. Due to the fact that such evaporators can also be used to maintain the pressure in tanks, it is more appropriate to install them during occasional operation of engines on natural gas (when fuel from tanks is consumed only for a while) rather than throughout the voyage. This will reduce the metal consumption and reliability of the system due to the absence of outboard heaters.

The disadvantage of submerged evaporators is complexity of the tanks design, incomplete use of fuel (because the coils must be necessarily covered with liquid), relatively small level of the tanks filling, whereby part of LNG evaporates directly in the tank.

Free space is necessary to prevent a dangerous increase in the pressure, since the volume of the vapor phase increases by hundreds of times compared with the liquid phase.

Tanks of a large volume should be placed in the hull of the vessel in a horizontal position because the filling line of the tanks is below the bunkering unit, the hydraulic losses are reduced, and as a consequence, the pump head and its drive power also decrease.

3.2 Main Equipment for LNG Fuel Systems of Ship Power Plants

Cryogenic pipelines and pumps, storage tanks, and heaters can be ascribed to such equipment.

Cryogenic pipelines are designed to transport liquid cryogenic products with a minimal heat inflow from the environment. They are double-walled, most often with a screen-vacuum insulation. Sections of the cryogenic pipelines are connected together by means of welded joints. Pipeline systems can be completed with the following elements: cryogenic pipes; elements of compensation; connecting elements; adsorption sections; vacuum ports for connection; means of vacuum processing and vacuum control; supports, membrane protective devices and protective earthing devices. Figure 3.8 shows the layout of the LNG bunker pipeline equipped with pressure sensors, ventilation system between the walls of the inner and outer housings, gas sensors, and drip tray [13, 14].

Bunkering of the LNG vessels can be carried out from specialized coastal or floating refueling complexes at specialized berths. As an exception, it is possible to refuel the vessel with LNG at the berths of general use, in specially designated places by the "direct option".

The coastal berths for LNG supply include isolated storage tanks for the gas coming from the NG liquefaction port facility, pipelines and fuel supply pumps. The berth accommodation project should provide the protection of the fuel distribution

Fig. 3.8 LNG bunkering pipeline: 1—stainless steel drip tray; 2—double-walled LNG feeder; 3—pipeline ventilation device; 4—gas sensors; 5—inner pipe for liquefied gas supply into the tanks

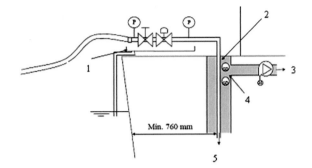

system from strikes of lightning, static electricity and electromagnetic induction, the possibility to receive the vessel signal to cut off the bunkering and automatically detach the LNG pipe or fuel distribution system. Connections of bunkering pipelines should be carried out taking into account the location of cryogenic tanks on the vessels of different types, lower and higher water level, change of the ship's freeboard and admission to drift.

Bunkering of the LNG vessel is to be carried out at daylight hours. If this condition is not feasible, an explosion-proof stationary or portable electric light should be provided. All flange connections including articulated joints must be earthed through the armor with an electric cable [4].

Bunkering of ships with liquefied natural gas can be carried out with the help of floating refueling complexes—special transport vessels (barges) on which LNG tanks are located, fuel supply is carried out by cryogenic pumps through isolated pipelines. The bunkering complex includes: LNG terminals, trucks for fuel transportation, tanker vessels, and ground cryogenic tanks [10, 14]. Figure 3.9 shows bunkering of the vessel from the tanker barge. The bunkering receiving terminals are located on both sides of the vessel. A sampling unit must be installed at the receiving equipment to connect sampling devices and product quality control devices. The frequency of bunkering depends on the range of the vessel's voyage, the number and duration of parking. On average, refueling occurs every 3–7 days; the duration of bunkering is 2–3 h [15].

The LNG cryogenic pumps consist of two main units:

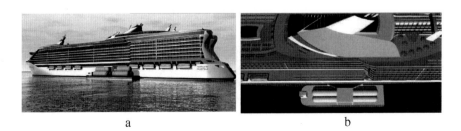

a   b

Fig. 3.9 Vessel bunkering from the floating refueling complex: **a** side view; **b** top view

- the pump for inlet and pumping of cryogenic liquids;
- the hydraulic drive motor or other type of drive.

The pump set is installed in the vertical position or at an angle. When the inlet pressure reaches the required value (depending on the characteristics of the pump), the cryogenic liquid is taken from the tank. The control valve prevents the liquid back flow.

Onboard of the ship, LNG is stored in securely protected tanks; they are usually cylinders with rounded ends. This form of tanks provides a reasonable ratio of mass, volume and cost. Of course, a spherical shape is optimal, but such tanks are very labor-intensive in manufacturing. To ensure the maximum possible volume with minimum heat input, the tank diameter/length ratio should be as large as possible, i.e. it is necessary to provide minimum surface area for a given volume.

LNG tanks usually consist of two chambers—internal and external, between which deep vacuum is maintained to isolate the liquefied gas, which is usually stored at 162 °C. The pressure within the vacuum cavity is $10^{-1} \ldots 10^{-2}$ Pa. The vacuum insulation is carried out in such a way as to provide reliable vacuum in the space between the tanks. Typically, the inner casing is made of stainless austenitic steel, nickel steel or composite materials. The outer casing is made of carbon or low-carbon steel in accordance with existing standards. Pipelines inside and outside the tank are made of stainless steel. Throttle valves within the inner casing of the tank minimize the oscillations of the LNG surface during the movement of the vessel carrying fuel tanks onboard [16–19].

The LNG tank should not be filled completely because there should be space for the liquid vapor, so the filling factor is 0.88–0.95. The tanks can be horizontal and vertical in design. The advantage of horizontal tanks is the fact that for the equal volume, the area of the liquid mirror is larger than in the vertical ones (the larger the area is, the more intensive evaporation process is). To ensure the minimum energy potential for fire and explosion hazard, the storage option with the least excess pressure is to be preferred.

Figure 3.10 shows the layout of the marine LNG tank together with the hung "cold box" where the evaporator and distribution valves are located [13, 14].

Evaporators of two basic types—of direct and indirect heating—are used for the LNG regasification.

The first group includes devices in which heat is supplied to liquefied gas through the wall directly from the hot coolant: coil, tubular, irrigation and fire. The second group includes the devices in which liquefied gas receives heat through the wall from an intermediate gas or liquid heated by the hot coolant. For example, there can be a fire evaporator with a water bath, in which the intermediate coolant is nitrogen or helium, and an electric evaporator, in which the intermediate coolant is nitrogen.

In addition, regasification units can be classified according to the following characteristics:

- regasification scheme implemented (capacitive, flow, combined);
- type of the coolant contact with liquefied gas (electric, fire, water, steam, oil);

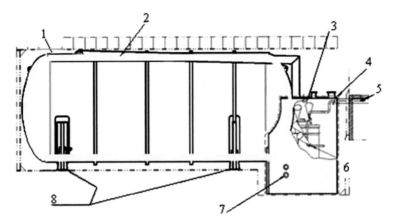

Fig. 3.10 Marine LNG storage tank: 1—tank outer shell; 2—LNG feed and vent lines; 3—gas sensors; 4—casing where the evaporator and fittings are installed; 5—gas supply pipeline into the engine room; 6—control armature; 7—evaporator (input/output); 8—flanges of the SPP liquefied natural gas

- contact of liquefied gas with the heating surface (with boiling of liquefied gases, boiling in pipes under forced circulation, irrigation—film and nozzle);
- evaporative capacity (small, medium, large) [20, 21].

LNG regasification may employ evaporators using different types of energy:

- atmospheric, with preheaters to provide the required outlet temperature applying the environmental heat to further preheat the gas up to the specified temperature, before the supply into the engine, the heater is installed next to the evaporator;
- using the heat of the thermal engine exhaust gases;
- with heating by hot water, steam and other coolants, including non-freezing ones;
- with electric heating.

To prevent freezing of the coolant within the evaporator, it begins to circulate earlier or at the same time as LNG enters. With a decrease in the specific fuel consumption per engine, the portion of the vaporized gas can be supplied into the intermediate (buffer) tank.

When LNG is regasified under shipboard conditions, the following options are possible: evaporation and heating may occur in one heat exchanger or in two (the liquid phase evaporator and gas preheater). In the first case, it is possible to use the directly heated hot-air evaporators-heaters, in particular the "pipe-in-pipe" design. The gas burner is installed under the inner tube (the cylindrical furnace). The combustion products move along the pipe, and the coil is located on the combustion chamber. Liquefied gas moves along the coil. The outer pipe is insulated. Individual evaporators can be submersible or remote. Submersible evaporators in the form of coils are mounted inside the tanks; they can be electric or with the steam coolant. The remote evaporators can be designed in the "shell-and-tube" form with tube bundles or in the "pipe-in-pipe" form. Hot water, "water–glycol" mixtures, water or vapor can be used as the coolant; alternatively, evaporators with heat-electric heating elements can be applied.

It is advisable to use the plate or plate-fin heat exchangers as the gas heaters. The heat exchangers should be compact with the minimum possible mass dimensions. It is important to consider a possible increase in the consumption of electrical energy (when using electric heaters), increase in the required capacity of steam or hot water boilers and evaporative installations (when using fresh hot water or steam), increase in the capacity of pump drives for pumping hot coolants (especially when using seawater), as well as the possibility of utilizing "cold" energy in ship conditions.

The use of recuperative heat exchangers is an acceptable option for LNG gasification onboard of the vessels. Evaporation of the NG liquid phase should occur within pipes of a large diameter, as the volume can increase up to 600 times. LNG gasification of is carried out in two stages: evaporation and transfer to the gaseous state, heating of the gaseous fuel to the required temperature about (0…60) °C. In this case, two heat exchangers can be installed: a liquid phase evaporator and a gaseous fuel heater.

Let us consider some schematic solutions of the regasification units and equipment for LNG evaporation.

Figure 3.11 shows the schematic diagram of the installation for LNG gasification using ethylene glycol and water as the heating medium. The unit consists of two parts, for evaporation of liquefied natural gas and utilization. In the latter, the cooled water gives low-potential energy to the refrigerant (ethylene glycol). Such diagrams are effective at a high specific consumption of the liquefied gas, and, as the consequence, a large amount of low-potential energy is formed from the heat exchange.

Fig. 3.11 Diagram of the cryogenic liquid gasification (US Patent No. 7155917): I—scheme of LNG evaporation with ethylene glycol and water coolants; II—system of coolant circulation; 1—evaporator; 2—heater; 3, 7—tanks; 4—circulating pump; 5—heat exchanger; 6—air heater; 8—circulating pump; 9—control valve block

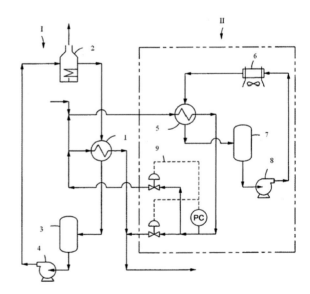

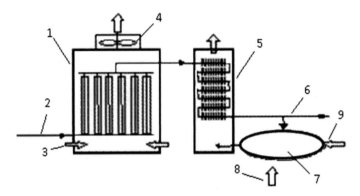

Fig. 3.12 LNG gasification within air evaporators: 1—evaporator; 2—LNG supply; 3—air; 4—fan; 5—superheater; 6—natural gas (gaseous state); 7—air heater; 8—hot air for LNG evaporation; 9—air for heating

Such circuit design can be rationally used in cogeneration plants with gas or gas-diesel engines (if the gas is stored in the liquefied form), as well as on passenger or cargo-and-passenger ships where there is a great need in "cold" energy for air conditioning and/or refrigeration units [22].

Air can be used as the coolant for LNG gasification (Fig. 3.12). Before being fed to the heat exchanger, the air is heated in the preheater, where part of the vaporized gas is used as fuel. To obtain dry gas without drops of the liquid phase, the natural gas after the air heater is directed to the superheater [23].

Let us consider the main types of regasification units used for LNG evaporation: with water irrigation, with submerged burner and intermediate coolant [19].

In evaporators with water irrigation (Fig. 3.13), LNG is distributed through the collector to the vertical pipes assembled in the mechanical core. The irrigation water flows along the height of the pipes; it is supplied to the drain pipes from above and collected downstream into the common collector [19, 24]. The water-LNG heat exchange is carried out according to the counter flow diagram (Fig. 3.13a) or the counter flow and direct-flow diagram (Fig. 3.13b). The pipe profile is finned to improve heat transfer. The scheme is simple, therefore operation and maintenance is

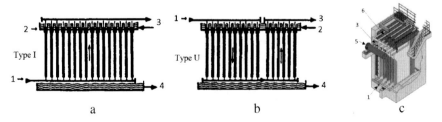

Fig. 3.13 Water irrigation regasification units: **a** with vertical pipes; **b** with U-shaped pipes; **c** general view; 1—LNG; 2—hot water; 3—natural gas; 4—cooling water; 5—sea water; 6—sea water distribution

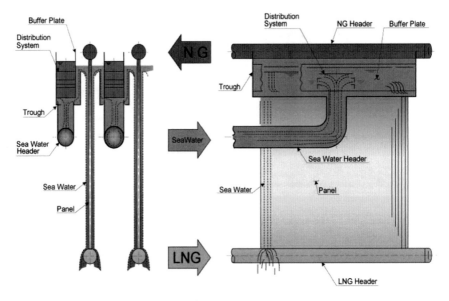

Fig. 3.14 Evaporation within water irrigation evaporators

also fairly simple. External inspection of the pipes is available. The heat exchange mechanism is stable. There is a possibility of processing large volumes of regasification by increasing the number of pipe assemblies. If seawater is used, it must be subject to additional treatment to prevent corrosion of pumps and water pipes [19].

The scheme of evaporation within the water irrigation regasification units is shown in Fig. 3.14 [24].

In gasification units with a submerged burner (Fig. 3.15), the tube bundle through which LNG circulates is submerged in a metal or concrete container filled with water [19, 25]. The submerged combustion burner is installed on the bottom of the tank and is supplied with gas and compressed air.

The gas flow from the burner is bubbled in the water layer, heating it up to ≈30 °C and mixing it, which ensures a good heat exchange between the water and the LNG and prevents formation of ice on the tube bundle. The regasification units of this type are compact enough considering the high efficiency of heat exchange between combustion products, water and LNG. Adjusting the gas-air mixture supply to the burner guarantees a small amount of nitrogen oxides in the combustion products. It is necessary to ensure regular neutralization of water by feeding the fresh portion of it, as water is acidified by the products of gas combustion and becomes corrosive [19].

The intermediate coolant regasification units (Fig. 3.16) are cryogenic heat exchangers with tube bundles that realize heat exchange according to the scheme "LNG—intermediate coolant—hot source". Such devices are well adapted to the processes of cold LNG recuperation, they are compact and highly efficient [19, 24].

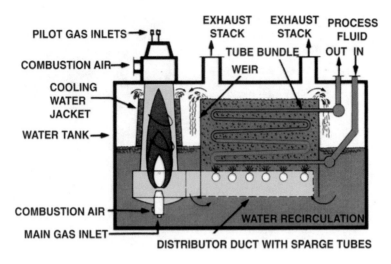

Fig. 3.15 Regasification unit with a submerged burner

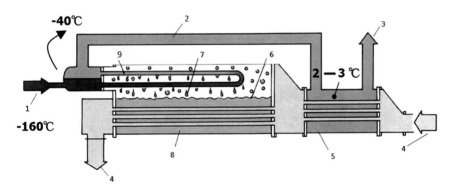

Fig. 3.16 Intermediate coolant (Fluid) regasification unit: 1—LNG; 2—LNG evaporator; 3—natural gas; 4—sea water; 5—natural gas heater; 6—intermediate coolant evaporation; 7—intermediate coolant condensation; 8—intermediate liquid coolant evaporator; 9—LNG evaporation

3.3 Parameters of LNG-Fueled Ship Power Plants

Liquefied natural gas belongs to the group of alternative fuels. If compared to traditional ones, it can change the type and composition of the propulsion complex, engine type, so the fuel system is changed radically. Efficiency evaluation is carried out when LNG is used as fuel for main engines and/or diesel generators.

Fuel consumption is one of the main indicators of the energy efficiency. Figure 3.17 presents the data on the amount of fuel consumed and its cost for the ferry between Hanko and Rostock. The data are presented for 5 options of the SPP package content [26].

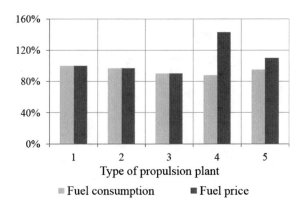

Fig. 3.17 Relative fuel consumption and price for the ferry SPP: 1—twin-screwdiesel-mechanicalpropulsion (DMP) onheavyfueloil (HFO) installed power 55.98 MW; 2—diesel-electric propulsion (DEP) on HFO installed power 53.1 MW; 3—diesel-electric-mechanical propulsion (DEMP) on HFO installed power 53.1 MW with a controllable pitch propeller; 4—DEMP on MDO installed power 53.1 MW with a controllable pitch propeller; 5—DEP on LNG installed power 51.9 MW with a controllable pitch propeller

As it can be seen from Fig. 3.17, the CPP ships have less fuel consumption due to lower power. However, the relative cost of fuel for DEMP on MDO is increasing, which is due to the higher price of MDO. The fuel cost for the DEP on LNG is slightly higher than for the HFO variants, since the price of gas in the calculations was taken by 40% higher than the cost of a ton of heavy fuel (as on 2009).

The efficiency coefficient of engines when operating on LNG is on average by 2…4.5% lower than for that for HFO [27]. Fuel consumption when operating on LNG is reduced by 5% compared to heavy fuel [10].

Mass and dimensional parameters of gas-diesel engines (GDE) practically do not differ from those for diesel engines. In most cases, they are a modification of standard diesel engines adapted to operate on gas fuel. The specific mass of the gas turbine engine is in the range of 2…32 kg/kW at the power of up to 5 MW. The range of values decreases on average to 14…25 kg/kW at a high power [28].

One of the significant drawbacks of using LNG on the vessel is a considerable increase in the mass and dimensions of the fuel tanks. Nevertheless, the combustion heat of LNG is higher than that of oil fuels, and less fuel is required (by 7…10%) [27]. Due to the considerable mass of double-hull fuel cisterns, which can account for up to 80% of the mass of fuel tanks, the mass values increase substantially.

The requirements of classification societies for the LNG fuel tanks placement on the vessel are as follows: the distance from the ship's side is not less than B/5 (but not less than 2 m), the distance from the ship's bottom is not less than B/15 (but not less than 760 mm), where B is the width of the vessel. This is necessary for the protection of containers in case of collision or vessel going aground, or in other cases where the hull may be damaged.

These requirements almost do not differ from the restrictions regarding the arrangement of diesel fuel tanks on newly built cruise liners. However, for LNG, these requirements are more associated with the significant volume of fuel tanks; at the atmospheric pressure in tanks, their volume is by 1.8 times larger than that for diesel fuel with the same energy intensity. The actual volume of storage space for LNG tanks is almost 4 times larger than that for MDO, taking into account the provision of empty space around cylindrical tanks. The restriction on the maximum volume leads to the change in the cruise ships structure; with the same passenger capacity, the PG cruise ship will have larger dimensions [7, 29].

On the drilling platforms servicing vessels, the engine room (MO) is usually divided into two separate zones, one of which contains engines, while the other contains tanks with LNG [30].

Figures 3.18, 3.19, 3.20 and 3.21 show the layout of the SPP main equipment, including the LNG storage tanks [10, 13, 29]. The layouts in Fig. 3.19, 3.20 and 3.21 show diesel generators operating on gas, the main engines operate on liquid fuel, cryogenic tanks are located horizontally. Cylindrical cryogenic tanks are delivered to the vessel in standard 40-foot containers in the form of parallelepiped (the main variant of placement). Figure 3.21 shows the diesel generators operating on gas, the main engines running on liquid fuel, cryogenic tanks located vertically.

Existing regulations establish the following requirements for the storage of natural gas as fuel on ships:

– in case of the tanks placement on the deck, the PG can be stored both in the liquefied and compressed state, the tray is installed under LNG tanks if there are connections below the liquid level in the tank;
– in case of the tanks placement below the deck (in the ship's hull), the maximum storage pressure is limited to 1 MPa, mainly LNG is used, and the tanks should be located in a special room. Stainless steel or other materials resistant to low temperatures are used for the tanks manufacturing. The tanks are designed for the maximum permitted storage pressure on the vessel, they are necessarily equipped with a pressure-maintaining valve. In case of the pressure increase, the portion of the gas is discharged through it. In all cases, there must be reliable isolation [31].

Figure 3.22 shows the cross sections of the belly of the vessel using heavy fuels and LNG. When LNG is applied, the use of double bottom tanks and the LG tanks storage in the vessel hold are not permitted. This is due to the need to comply with fire and explosion safety requirements in case of methane leakage or tanks damage in case of an accident. Another reason is the presence of a large number of fittings (including those for ventilation and maintenance of vacuum in the tank), which complicates maintenance by minimizing the length of LG and gaseous fuel transport pipelines [15].

Fuel tanks with LNG can be located vertically in steel casings within the ship superstructure, single-tier horizontally in the engine room (in one plane with the engines) and two-tier (on the special platform above the engines), with the small-volume tanks—directly above the engine vertically in the steel casing (Fig. 3.23).

The basic elements of the layouts in Fig. 3.23c and d are the bunkering station through which fuel tanks are filled, the LNG storage tanks, and the "cold box" where evaporator heaters, fuel system valves (gas valves) and bi-fuel engines are located

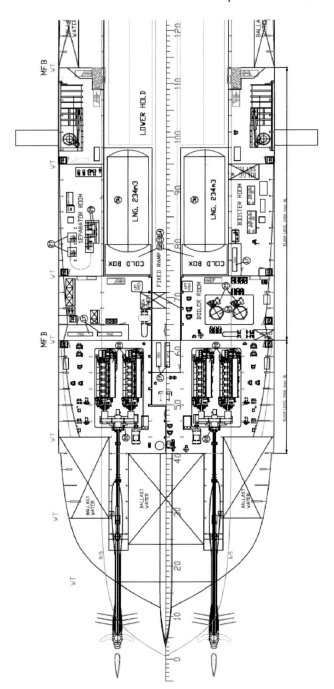

Fig. 3.18 Equipment layout within the ship engine room with the main gas-diesel engines

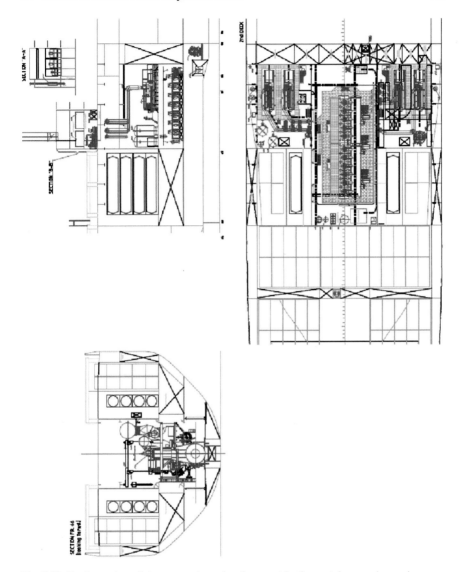

Fig. 3.19 Basic version of the cryogenic tanks placement in the container carrier engine room (Eight 40-foot containers of 31.5 m^3 with total capacity of 250 m^3, volume occupied is equivalent to 40 TEU)

within the metal casing. In the basic version (Fig. 3.23c), the tanks are located in the same plane as the engines and the fuel supply pipelines. The alternative version involves placement of the cryogenic tanks and "cold boxes" above the engines, that is, the two-tier arrangement of equipment in the engine room [15].

One of the reasons for using LNG on passenger cruise ships is that natural gas is the sufficiently safe fuel; its flash point is notlower than 135 °C. The self-ignition

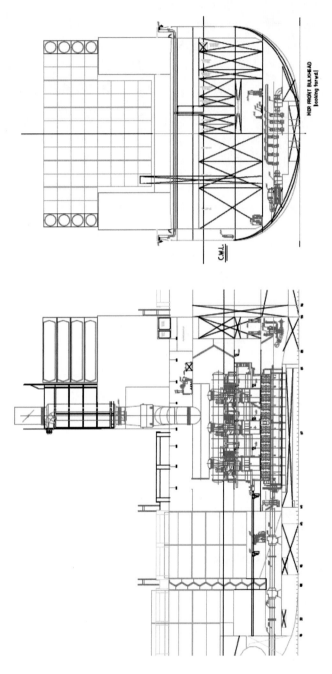

Fig. 3.20 Alternative version of the cryogenic tanks placement on the container carrier deck (Eight 40-foot containers of 31.5 m³, total capacity of 250 m³, volume occupied is equivalent to 16 TEU)

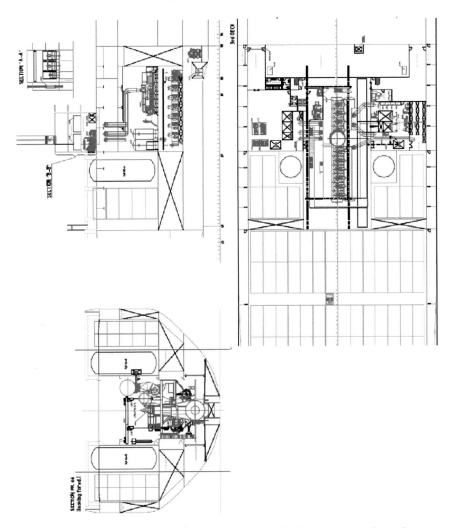

Fig. 3.21 Layout of the cryogenic tanks vertical placement in the container carrier engine room (two vertical tanks with capacity of 190 m^3, volume occupied is equivalent to 20 TEU)

temperature is 537 °C, which is much higher than similar values for oil fuels. When gas is stored in the liquefied state at very low temperatures, the probability of self-ignition is practically zero. The tanks with liquefied gas are not to be placed in holds. The explosive gas concentration in the room is 5–15% of the air volume, for diesel fuel—2...3%. In case of possible leaks, the gas does not accumulate in the room because it is lighter than air. Its leakage can be determined quite quickly with special equipment, which must be necessarily mounted in the engine room (ER) [4, 32]. Automated systems for gas distribution to consumers and detection of gas leaks must be installed in the ER (gas detectors allow detecting even small leakage).

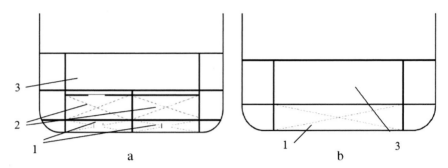

Fig. 3.22 Cross sections of the vessel belly: **a** SPP operation on heavy fuel oil; **b** SPP operation on LNG; 1 double bottom; 2 heavy fuel tank; 3 lower hold

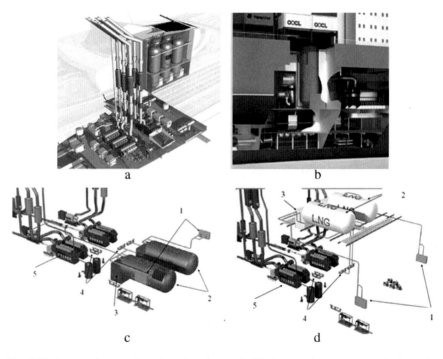

Fig. 3.23 Layout of cryogenic tanks onboard: **a** vertical in the vessel superstructure; **b** above the engine; **c, d** in the engine room (basic and alternative versions); 1—bunkering station; 2—cryogenic tanks; 3—evaporator; 4—gas supply valves; 5—bi-fuel engines

Double-walled pipelines are used for the NG supply after the evaporators. A large amount of natural gas is not allowed in the engine room; the room should be well ventilated. As a rule, LNG is stored in special containers in separate rooms, which are at the same level as the passenger cabins above the ER [7, 10, 13]. In cargo-passenger ferries, the tanks with LNG are also located in the separate room, but unlike cruise liners, it is located in the same plane and next to the ER [10, 33].

Compared to other types of ships operating on gas, cruise ships employ a different concept of the LNG tanks arrangement. The tanks are located in the center of the superstructure between the outer rows of cabins in front of the ER. The fuel tanks are placed above the decks in order not to interfere the passenger flow. Quite often technical rooms of modern cruise ships, for example, for air conditioning, are located in superstructures that do not have natural lighting. The room for fuel tanks is a logical extension of these technical premises. The free access of air is the advantage of this arrangement, which ensures sufficiently reliable storage. Any possible leaks of natural gas evaporate and dissipate in the air. The steel tray under the tanks is needed to collect condensate and prevent the contact of liquefied gas with the hull parts in case of possible leaks. The tanks are placed in the special casing necessary for fire protection [9–11].

The emissions of carbon dioxide and nitrogen oxides are significantly reduced; there are no sulphur emissions as the result of the LNG use on the vessel (Fig. 3.24).

The data are given for high-speed vessels of the Ro-Pax type. Catalytic neutralization systems are provided in the propulsion systems of types 2–4. It provides a low level of nitrogen oxide emissions for these options [26]. The highest level of emissions is typical for DMP on HFO. On average, when using LNG compared to heavy fuel, CO_2 emissions are reduced up to 30% due to the low ratio of carbon to hydrogen, NO_x is reduced up to 85% due to combustion of the lean mixture, the level of particulate emissions is reduced up to 99%, and there is no visible smokiness [34].

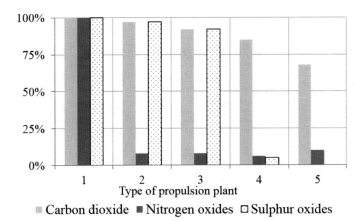

■ Carbon dioxide ■ Nitrogen oxides □ Sulphur oxides

Fig. 3.24 Relative level of the vessel emissions: 1—twin-screw diesel-mechanical propulsion (DMP) on heavy fuel oil (HFO) installed power 55.98 MW; 2—diesel-electric propulsion (DEP) on HFO installed power 53.1 MW; 3—diesel-electric-mechanical propulsion (DEMP) on HFO installed power 53.1 MW with a controllable pitch propeller; 4—DEMP on MDO installed power 53.1 MW with a controllable pitch propeller; 5—DEP on LNG installed power 51.9 MW with a controllable pitch propeller

Due to the fact that the most severe restrictions on ship emissions are introduced in coastal areas of navigation, it is of interest to use natural gas as the fuel when vessels are in these areas, as well as when they are parked. Usually engines of the ship power plant operate on heavy fuels (both main and diesel generators).

However, when the vessel enters the areas where severe emissions restrictions operate, the main engine can be shut down, and energy requirements (including propulsion drives) are considered only with diesel generators that operate on natural gas. Several container carriers have already been operating in this operation mode.

Figure 3.25 provides the comparison of the annual operating costs for the ferry SPP [26]. The lowest annual costs are for DEMP on the HFO, while the costs for DMP on the HFO are slightly higher. The cost of the DEP on HFO and LNG is higher because of the high cost of the electric propulsion systems. When the vessel operates on LNG, there is no need for catalytic purification of exhaust gases. According to various sources, the operating costs for the voyage fuel consumption when using HFO and LNG are lower by 18…24% for gas [10, 27].

The capital investments in the construction of ships with ship power plants operating on NG can increase by up to 30% compared to the use of HFO. The main increased expense items are the cost of the GDE and LNG fuel system. The annual

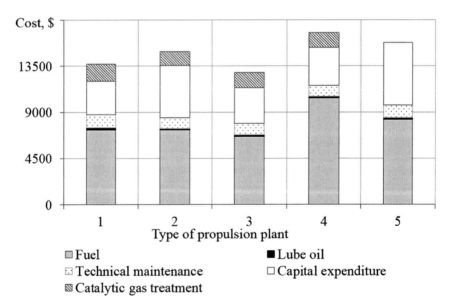

Fig. 3.25 Comparing the SPP Annual Fuel Consumption*, thousand USD: 1—twin-screw diesel-mechanical propulsion (DMP) on heavy fuel oil (HFO) installed power 55.98 MW; 2—diesel-electric propulsion (DEP) on HFO installed power 53.1 MW; 3—diesel-electric-mechanical propulsion (DEMP) on HFO installed power 53.1 MW with controllable pitch propeller; 4—DEMP on MDO installed power 53.1 MW with a controllable pitch propeller; 5—DEP on LNG installed power 51.9 MW with controllable pitch propeller * when the annual capital expenditures were calculated, the loan period was 10 years, and the annual loan rate was 8%

operating costs for the same vessel will be by 7% lower when using natural gas, the main reduced expense item is the fuel cost [10].

The cost of electricity generation for the vessel staying in the port is the lowest when diesel generators operate on LNG and makes up 0.06 USD/kWh. As a comparison, the cost of electricity from near-shore sources is 0.09 USD/kWh, with the engine running on MDO—0.11 USD/(kWh), on gas oil—0.13 USD/kWh [10]. The low cost of electricity generation when using LNG can be caused not only by the large index of the net calorific value, but also by various tax discounts and benefits that are granted to producers who use alternative fuels.

The range of the GDE operation is quite wide; its use allows maintaining the maneuverability of the vessel. For example, when the GDEs are used as diesel generators, the load on engines is 80% during parking, 85% during maneuvering, and 88% directly during the voyage at the total installed capacity of 15.04 MW. Similar values for diesel generators on MDO are 78%, 88% and 86% respectively, with the total capacity being equal to 14.4 MW [10].

When choosing the type of the vessel's operation on LNG, it is necessary to take into account that gas-diesel engines are commonly used as part of the diesel-electric propulsion unit (DEPU).

Figure 3.26 shows the SPP diagram with four gas-diesel engines. Two engines transmit torque through the gears to the screws and two shaft generators. Other engines drive electric generators. Part of the electricity generated by diesel generators and shaft generators is delivered to the rotorcraft complex, and the other is intended to the general needs [13].

There are separate data indicating that when LNG is used, the autonomy indicators are deteriorating. This is due to the significant mass and dimensional values of natural gas fuel tanks. At the same time, while maintaining the same size of the vessel, the mass of LNG fuel stock is smaller than that for petroleum fuel.

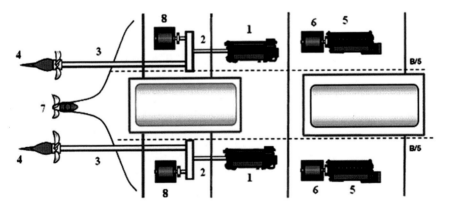

Fig. 3.26 Structural diagram of the SPP with gas-diesel engines: 1—main engines; 2—reduction gear; 3—shafting; 4—screw; 5—powered diesel engines; 6—power generators; 7—propelling motor; 8—shaft generators

3.4 Conclusions

1. The analysis of the existing schematic design made it possible to identify the main requirements for the SPP fuel systems on LNG taking into account the ship conditions. The recommendations on the specific features of fuel systems for liquefied natural gas can be used during selection and calculation of individual elements and systems.
2. The basic diagrams of the LNG fuel systems are presented. They differ in placement, quantity and arrangement of tanks (vertical or horizontal), quantity and type of evaporators-heaters, and organization of the regasification process.
3. LNG is the alternative to both diesel and heavy oil fuels. It can be used in both main engines and diesel generators. The promising segment of vessels for the LNG use are voyage vessels that navigate mainly in areas with severe environmental restrictions where there is a possibility of LNG bunkering, since not all port terminals have available terminal units.

References

1. V. M. Gorbov, V. S. Mitenkova, "Osobennosti toplivnyih sistem SEU na prirodnom gaze [Characteristic properties of ship power plants fuels systems using liquefied natural gas] ", Aviatsionno-kosmicheskaya tehnika i tehnologiya, pp. 20–24, #7, 2008.
2. J. Wegrzyn (2003) LNG fuel systems technology on-board: LNG pumps, storage tanks, and heat exchangers [Online]. Available: https://www.afdc.energy.gov/afdc/pdfs/lng_fuel_sys_tech_w egrzyn.pdf.
3. D. Zbaraza (2004) Natural gas use for on-sea transport [Online]. Available: https://www.ipt. ntnu.no/~jsg/studenter/diplom/DawidZbaraza2004.pdf.
4. Pravila bezopasnosti pri proektirovanii i ekspluatatsii sistem priema, hraneniya, zapravki i gazifikatsii szhizhennogo prirodnogo gaza na ob'ektah potrebitelya [Safety rules in the design and operation of handling, storage, bunkering and gasification systems on consumer's objects], Russian Gosgortehnadzor PB 08–342–00, 2001.
5. S. Iwai, S. Sugiyama, Y. Yamasaki, and others (2003) Plan for the installation of a power plant using LNG cold energy at an LNG terminal [Online]. Available: https://www.igu.org/html/wgc 2003/WGC_pdffiles/10391_1045242396_18621_1.pdf.
6. "Rising to the regas challenge", The Naval Architect, pp. 24–26, March 2009.
7. M. Berisa, "New cruise ship concept from Wartsila (Part 1)", Twentyfour7, pp. 51–54, #2, 2007.
8. "Clean Cruising Concept", Diesel & Gas Turbine Worldwide, pp. 59–61, Apr. 2008.
9. O. Levander, "Cruising on gas into a cleaner future", Wärtsilä technical journal, pp. 26–31, #1, 2007.
10. O. Levander (2008) Turning the page in ship propulsion, by switching to LNG [Online]. Available: https://www.dieselduck.ca/library/05%20environmental/2008%20Wartsila%20propuls ion%20alternatives.pdf.
11. Wärtsilä engine products (2010) [Online]. Available: https://ip2010.turkuamk.fi/pdf/Wartsila_ engine_products.pdf.
12. V. M. Gorbov and V. S. Mitenkova, "Palyvna systema zberihannya ta pidhotovky zridzhenoho pryrodnoho hazu [Fuel system for natural gas storage and treatment]", Ukraine Patent 50594, June 10, 2010.

13. T. G. Osberg (2008) Gas engine propulsion in ships. Safety Considerations [Online]. Available: https://www.bi.edu/ShippingakademietFiles/IBC/Presentations/Thursday/Torill%20Grimstad%20Osberg.pdf.

14. T. G. Osberg (2011) Gas fuelled engine installations in ferries Online]. Available: https://www.interferry.com/2011papers/3-1Presentation_Osberg.pdf.

15. J. Paananen, LNG fuelled ships and auxiliaries. Rostock: Wärtsilä ship power Publ., 2007, 44 p.

16. M. G. Kaganer, Teplovaya izolyatsiya v tehnike nizkih temperatur [Heat insulation in cryoengineering]. Moskva: Mashinostroenie Publ., 1966, 275 p.

17. E. I. Mikulin, Kriogennaya tehnika [Cryoengineering]. Moskva: Mashinostroenie Publ., 1969, 272 p.

18. M. P. Malkov, I. B. Danilov, A. G. Zeldovich, A. B. Fradkov, Spravochnik po fiziko-tehnicheskim osnovam kriogeniki [Guiede of physicotechnical fundamentals of cryoengineering]. Moskva: Energoatomizdat Publ., 1985, 432 p.

19. Entsiklopediya gazovoy promyishlennosti [Gas industry encyclopedia] under the editorship K. S. Basnieva. Moskva: TVANT Publ., 1994, 884 p.

20. N. L. Staskevich, D. Ya. Vigdorchik, Spravochnik po szhizhennyim uglevodorodnyim gazam [Liquefied hydrocarbon gases guide]. Leningrad: Nedra Publ., 1986, 543 p.

21. N. L. Staskevich, G. N. Severinets, D. Ya. Vigdorchik, Spravochnik po gazosnabzheniyu i ispolzovaniyu gaza [Gas supply and consumption guide]. Leningrad: Nedra Publ., 1990, 762 p.

22. Ned P. Baudat, "Apparatus and methods for converting a cryogenic fluid into gas", US Patent 7155917, Feb. 1, 2007.

23. Hot air draft super heater with air fin LNG vaporizer for satellite stations (2012) [Online]. Available: https://www.tokyo-gas.co.jp/techno/stp/02a9_e.html.

24. M. Endo (2013) Overview of ORV and IFV characteristics and operation in LNG receiving terminals in Japan and worldwide [Online]. Available: www.lngexpress.com/rcp/presentations/MasaoEndo110905.pdf.

25. Vaporizer alternatives study. Oregon LNG import terminal (2007) [Online]. Available: https://s3-us-west-2.amazonaws.com/oregonlng/htdocs/pdfs/appendices/RR_13-1/Appendix13R-p2.pdf.

26. O. Levander (2002) Advanced machinery with CRP propulsion for fast RoPax vessels [Online]. Available: https://www.wartsila.com/Wartsila/global/docs/en/ship_power/media_publications/technical_papers/advanced_machinery_with_crp_propulsion_for_fast_ropax_vessels.pdf.

27. V. M. Gorbov, Yu. A. Shapovalov, G. G. Kirchev, Vyibor pokazateley energeticheskoy ustanovki, harakteristik SEU i sudna na nachalnyih stadiyah proektirovaniya sudov transportnogo flota [Choise of ship power plant and ship parameters on the stage of conceptual design for transport vessels]. Nikolaev: UGMTU Publ, 1997, 141 p.

28. V. M. Gorbov, V. S. Mitenkova, "Analiz perspektiv ispolzovaniya prirodnogo gaza v sudovyih dvigatelyah vnutrennego sgoraniya [Analysis of natural gas usage perspectives in marine internal combustion engines]", in Issledovanie, proektirovanie i ekspluatatsiya sudovyih DVS, Sankt-peterburg, 2008, pp. 105–110.

29. O. Levander, "Environmentally-sound cruising", Twentyfour7 (Wartsila), pp. 48–50, #1, 2006.

30. M. K. Sandaker (2008) Use of natural gas as fuel for ships [Electronic resource] [Online]. Available: https://www.uscg.mil/marine_event/docs/panel1_doc1.pdf

31. P. Einang (2014) LNG som drivstoff for skip [Online]. Available: https://www.gasforeningen.se/upload/files/seminarier/gasdagarna2007/foredrag/per%20magne%20eingang.pdf.

32. V. A. Zagoruchenko, A. M. Zhuravlev, Teplofizicheskie svoystva gazoobraznogo i zhidkogo metana [Thermophysical properties of gas and liquid methane]. Moskva: Izdatelstvo standartov Publ, 1969, 236 p.

33. T. G. Osberg (2008) Gas Fuelled Engine Installations in Ferries. Emission Reductions and Safety Considerations [Online]. Available: https://www.sname.org/HigherLogic/System/Dow nloadDocumentFile.ashx?DocumentFileKey=be40b7e0-654f-44f4-a3d1-b1a8fa293c9b.
34. "LNG-fueled engines for offshore vessels", The Naval Architect, pp. 26, Jan. 2002.

Chapter 4
Biodiesel and Its Blends as Marine Fuels

4.1 Review of Fuel Systems for Biodiesel-Fueled Power Plants

The main feature of the biodiesel (BD) fuel systems in comparison to other alternative fuels is the fact that their configuration and equipment are the same as for diesel fuel. If the system operating on water-fuel emulsion is to be equipped with additional devices for preparation of persistent emulsions, and the layout and composition of the equipment for LNG are fundamentally different from the petroleum fuel ones, then biodiesel can be used in standard systems designed for diesel fuel, as the physicochemical properties of these fuels are quite close.

At the same time, for the efficient use of biodiesel fuels taking into account their specific features, it is advisable to introduce a number of changes in fuel systems. For example, it can be the use of other materials, coatings, and basic equipment designed specifically for BD (tanks, filters, separators, pumps). The latter will differ from similar equipment for diesel fuels only by certain technical characteristics and materials used, but not by its arrangement. Before listing the recommendations on how to take into account the features of the BD fuel systems, it is necessary to analyze its characteristics in order to link them to specific operating conditions. The analysis of the main characteristics of biodiesel fuels was carried out in Chap. 2.

The diagram of the fuel system of the ship power plant (SPP) with a diesel engine, where heavy oil is used as the main fuel, and diesel or biodiesel is the auxiliary one, is shown in Fig. 4.1 [1]. The scheme of the system offered by Wärtsilä shows that configuration of the fuel systems and composition of the basic equipment for DF and BD are similar.

A number of recommendations on the BD use in the fuel systems of SPP should be adhered to for the system's effective operation. As a result of the performed analysis of the operational parameters of biodiesel fuels, and also taking into account the experience of its use, recommendations have been developed for the use of this fuel on ships. They are presented below.

© Shanghai Scientific and Technical Publishers 2021
X. Yang et al., *Alternative Fuels in Ship Power Plants*,
https://doi.org/10.1007/978-981-33-4850-9_4

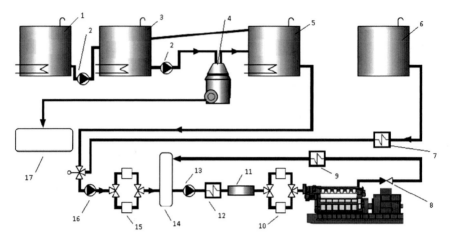

Fig. 4.1 SPP fuel system with the possibility of BD usage: 1—receiving tank; 2—fuel pump; 3—storage tank; 4—separator; 5—supply tank; 6—return fuel tank; 7—return fuel heater; 8—safety valve; 9—return fuel cooler; 10—fine filter; 11—viscometer; 12—heater; 13—booster pump; 14—mixing station; 15—filter; 16—fuel pump; 17—slurry tank

1. It is necessary to take the fuel samples regularly to check its quality and use BD with characteristics that meet the world standards. Engine manufacturers must provide quality requirements for B100 if it is intended for use in engines. Such requirements have already been developed and presented by such engine-building companies as Cummins, John Deere, MAN, Caterpillar, etc.

2. It is not recommended to store biodiesel for a long time under the direct impact of high or low temperatures; the storage life of fuel blends with diesel fuel depends on the BD concentration. The fuel should be stored in clean and dry containers within the dark room. It is necessary to remove water from the tanks before filling them with BD. To minimize moisture condensation during storage, it is necessary to keep the fuel tanks as full as possible, to drain and wash them before and after storage of biodiesel. Fuel tanks must be sealed to prevent water entering. It is necessary to clean the drain pipes regularly to prevent corrosion when using blends with diesel fuel at the BD concentration above 20%, as well as to monitor the water content and presence of microorganisms in the fuel regularly.

3. It is not recommended to store tissue materials moistened in BD for a long time. They begin to decompose with the release of heat, which can lead to spontaneous combustion.

4. At low temperatures, it is recommended to heat BD in tanks, pipelines, filters or use DF or its blends with the low BD concentration for the engine. It is necessary to implement chemical treatment with additives for normal fuel use at low temperatures. It should be taken into account that standard additives that reduce the pour point of fuels of the oil origin are not always suitable for BD. B100 should be stored at the temperature of (3...6) °C above the cloud point.

It is also possible to store BD at the temperature of (4.5...7) °C; depending on the raw material, this value can be higher. Tanks with fuel are to be insulated or heated.

5. If the engine operates on the mixture of BD and DF, it is recommended to use the flow homogenization method to obtain the stable and uniform fuel mixture.

6. The use of BD in small amounts (up to 2%) as the additive to diesel fuel with the low (.5%) and ultra-low (.05%) sulfur content improves the lubricating properties of diesel fuel. As a result, the fuel equipment wear decreases.

7. The experience of the B100 use for transport vehicles shows that biodiesel can be stored for 2–4 months without loss of stability. The ASTM data show that B100 can be stored with minimum stability for up to 8 months, with maximum stability—for one year or more. It is indicated that a week of fuel storage at the temperature of +43 °C is equivalent to a month of its storage at +21 °C. On average, it is recommended to store B100 for up to 6 months.

8. The above information provides an opportunity to formulate recommendations which would allow determining the conditions that correspond to the high level of the biodiesel fuel stability:

   – the higher the level of unsaturated compounds is, the lower the stability of the fuel is. Heat and sunlight can accelerate this process because it is undesirable to store the fuel under direct sunlight during the warm season;
   – such metals as copper, brass, bronze, lead, tin and zinc can also lead to an increase in the decomposition rate and formation of a considerable amount of deposits. B100 should not be stored for a long time in the systems that contain these materials;
   – some initial components of the interesterification process can wash away the natural antioxidants present in BD and thus reduce the stability of the fuel. Clarifying, deodorizing or distilling fats and oils before or during the biodiesel production also results in the removal of natural antioxidants;
   – the oxygen binding in the fuel reduces or eliminates the probability of the fuel oxidation and increases the fuel storage time;
   – antioxidants (natural ones or additives) significantly improve the fuel stability or its storage time without changing its properties.

9. When making the transition to biodiesel, it is necessary to remove sediments in the fuel system and equipment. When using blends with a high BD content, it is recommended to use filters with filter materials resistant to this fuel.

10. When using BD separately and with a high concentration of it in blends with DF, the tanks, pipelines and other elements of the fuel system in contact with it must have a protective coating. It is not recommended to use India rubber and natural rubber materials in fuel systems, or it is necessary to prevent fuel contact with the elements made of these materials. Teflon, Viton, fluorinated plastics and nylon are well compatible with B100, and the incompatible parts of the fuel system should be replaced. It is necessary to conduct regular inspection of the fuel system and equipment to detect leaks, seepage and softening of the

materials and to remove any spills of BD or its blends. B100 can be stored in the same fuel tanks as diesel fuel. However, it is more efficient to store biodiesel fuel in containers with a nitride coating, made of aluminum, stainless steel, fluorinated polyethylene and polypropylene, Teflon or fiberglass.

11. BD can be used in the same schemes and with the same equipment as DF, taking into account the specific characteristics of the former. In this regard, different options are possible. According to one of them, one universal fuel system for DF, BD and their blends in different proportions is provided on the ship. This is most suitable for small vessels with rigid mass-dimensional restrictions, provided that the fuels are used on the ship alternately and the ready-made blends of DF and BD are received without preparing them on board. By analogy with the use of heavy and light oil fuels, the other option suggests using two fuel systems with mutual reservation of individual equipment and preparation of fuel blends directly on the ship. Both options have their advantages, disadvantages, and the sphere of the most effective use [2].

Figures 4.2, 4.3, 4.4, 4.5, 4.6, 4.7 and 4.8 show possible options for the biodiesel fuel systems configuration and assembly. The diagrams have been developed on the

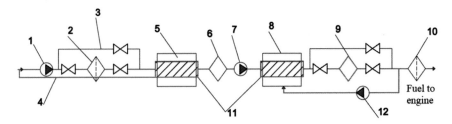

Fig. 4.2 Principal universal diagram of the fuel system for diesel and biodiesel fuels with external heating of tanks: 1—fuel pump; 2—coarse filter; 3—bypass line of the receiving pipeline; 4—satellite heater; 5—reserve fuel tank; 6, 9—separators; 7—fuel charger; 8—supply tank; 10—fine filter; 11—fuel heater in the tanks with external heating; 12—separator pump

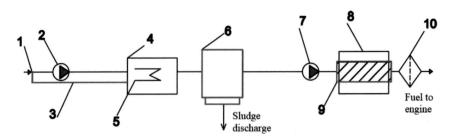

Fig. 4.3 Principal universal diagram of the fuel system for diesel and biodiesel fuels with combined filter-separator: 1—receiving pipeline; 2—fuel pump; 3—satellite heater; 4—reserve fuel tank; 5—submerged fuel heater; 6—combined fuel purification unit (filter-separator); 7—fuel charger; 8—supply tank; 9—fuel heaters in the tanks with external heating; 10—fine filter

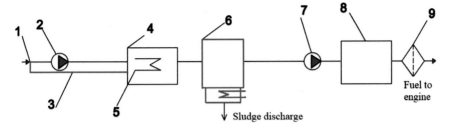

Fig. 4.4 Principal universal diagram of fuel system for DF and BD with combined filter-separator and built-in fuel heater: 1—receiving pipeline; 2—fuel pump; 3—satellite heater; 4—reserve fuel tank; 5—submerged fuel heater; 6—combined fuel purification unit (filter-separator with the built-in fuel heater); 7—make-up pump; 8—supply tank; 9—fine filter

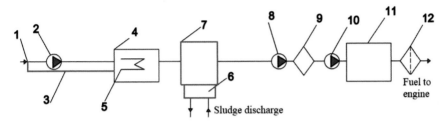

Fig. 4.5 Principal universal diagram of the fuel system for diesel and biodiesel fuels with the combined fuel filter-heater: 1—receiving pipeline; 2—fuel pump; 3—satellite heater; 4—reserve fuel tank; 5—submerged fuel heater; 6—fuel heater; 7—filter; 8—separator pumps; 9—separator; 10—fuel charger; 11—waste tank; 12—fine filter

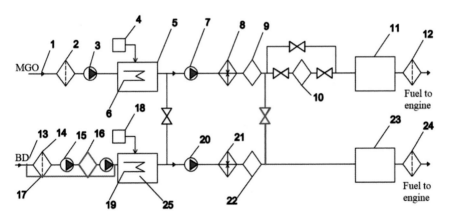

Fig. 4.6 Principal diagram of the diesel unit's fuel system suitable for operation on diesel and biodiesel fuels and their blends: 1—diesel fuel receiving pipeline; 2, 14—fine filter; 3, 8—fuel pumps; 4, 19—additive tanks; 5, 21—storage tanks; 6, 20—submerged fuel heaters; 7, 15, 22—separator pumps; 8, 23—heaters; 9, 24—separators; 10—mixer; 11, 25—supply tanks; 12, 26—fine filters; 13—biodiesel fuel receiving pipeline; 16—additional biodiesel separator; 17—satellite heater

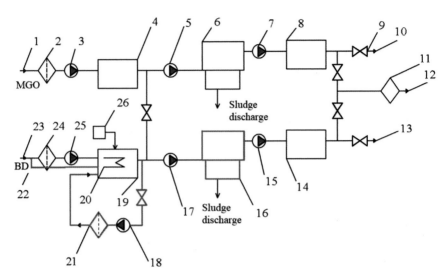

Fig. 4.7 Principal diagram of the diesel unit's fuel system suitable for operation on diesel and biodiesel fuels and their blends with combined fuel purification units: 1—diesel fuel receiving pipeline; 2, 24—coarse filters; 3, 25—fuel pumps; 4, 19—fuel storage tanks; 5, 17—pumps of fuel purification units; 6, 16—combined fuel purification unit (filter-separator with the built-in fuel heater); 7, 15—fuel supply pumps of supply tanks 8 and 14; 9—check valve; 10—diesel fuel supply to the engine; 11—mixer; 12—DF and BD blend supply to the engine; 13—BD supply to the engine; 18—separator pump; 20—submerged fuel heater; 21—separator; 22—satellite heater; 23—BD receiving pipeline; 26—BD additive tank

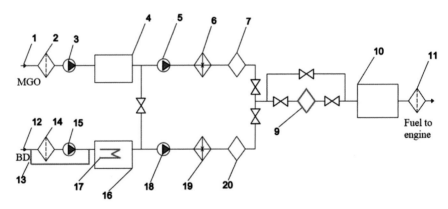

Fig. 4.8 Principal diagram of the diesel unit's fuel system suitable for operation on diesel and biodiesel fuels and their blends with the common storage tank: 1—diesel fuel receiving pipeline; 2, 14—coarse filters; 3, 15—fuel pumps; 4, 16—storage tanks; 5, 18—separator pumps; 6, 19—fuel heater; 7, 20—separators; 9—mixer; 10—charge tank; 11—fine filter; 12—BD receiving pipeline; 13—satellite heater; 17—submerged fuel heater

basis of existing ship systems for oil fuels taking into account the BD characteristics. When developing the options, it was taken into account that BD is the fuel intended primarily for the small vessels using light oil fuels. The diagrams in Figs. 4.2, 4.3, 4.4 and 4.5 provide the alternate use of biodiesel and diesel fuels, or their blends prepared before taking on board. Figures 4.6, 4.7 and 4.8 show two-fuel systems with the possibility of obtaining blends on the vessel and with reservation of separate equipment lines for DF and BD [3].

The diagram in Fig. 4.2 can be used with relatively small fuel reserves for the voyage and prolonged operation of the vessel at low ambient temperatures. The receiving pipeline is equipped with the satellite heater designed for heating B100 or blends with the high BD concentration (from 30 … 35%). The system includes the coarse filter. External heaters with electric heating elements are provided to heat the fuel within the storage and supply tanks. The system uses two separators—before and after the supply tank; a return fuel line is provided after the second separator in the supply tank. Due to a relatively high hygroscopicity of biodiesel, this solution will allow removing excess moisture from this fuel. When using DF and its blends with a low BD content, it is sufficient to have the separator after the fuel storage tank. There a bypass pipeline of fuel supply directly to the fine filter is provided after the fuel supply tank [3].

The special feature of the diagrams shown in Figs. 4.3, 4.4 and 4.5 is the availability of combined equipment. There, instead of two or three separate units, combined systems are used: filter-separator (Fig. 4.3), filter-separator with heater (Fig. 4.4), filter with heater (Fig. 4.5) [3].

These rather compact and efficient units are implemented at a relatively low fuel consumption, they use materials compatible with biodiesel fuel, and they are suitable for purification and heating of DF, BD and their blends as well. The use of such equipment makes it possible to reduce the mass-dimensional parameters of fuel systems and simplify their maintenance. The diagram in Fig. 4.3 shows the submerged coil heater installed in the reserve fuel tank and the built-in heater in the supply tank. In Figs. 4.4 and 4.5 there are no heaters within the supply tanks, since they receive heated fuel from the combined units. Such solutions are advisable at a small capacity of the supply tanks. Under operation at temperatures below zero, it is not recommended to use pure BD and blends with its high content in these diagrams. The separator is installed after the filter with the heater in the diagram in Fig. 4.5 [3].

Figure 4.6 shows that in addition to the coarse filter, the receiving line for biodiesel is fitted with the separator to take care of substantial watering. The mixer is located in front of the supply tanks on the diesel fuel supply line, which ensures reduction in the pump power capacity.

Ships predominantly employ the blends with a low BD content, so it is advisable to pump biodiesel fuel and not DF into the mixer. The diagram makes provision for separate tanks for diesel fuel and B100 [3].

The special feature of the diagram in Fig. 4.7 is the use of the combined treatment plants with the built-in heater during the fuel preparation; filtration and separation take place there [4]. The mixer is located after the supply tanks. There are three lines of fuel supply to the engines: for diesel, biodiesel and their blends.

In the B100 reserve tank, the circulating separator is provided for periodic removal of water from the fuel at its long-term storage (water may already be in the tank, or it may enter in a variety of ways during storage).

The diagram in Fig. 4.8 provides one supply tank for all the fuels; the mixer is installed in front of it. The bypass pipeline is provided to enable the engine operation on pure diesel fuel or B100. In addition to purification, mixing and heating of fuels in the systems using BD, chemical preparation should also be provided [3].

4.2   Main Equipment and Additives for Biodiesel Fuel Systems of Ship Power Plants

Additives. An effective method of the fuel performance improvement is its chemical treatment by adding additives. In the records, there is a lot of information on additives for heavy and light oil fuels that are used today for ships.

The analysis of the performance characteristics of biodiesel fuels made it possible to determine the main types of additives that can be used to ensure the normal functioning of ship power plants. For BD, the following types of additives are recommended [5, 6]:

- antimicrobial additives (the fuel is methyl or ethyl esters of fatty acids, which are a sufficiently favorable environment for the microorganisms' development);
- antioxidants (BD is oxidized fairly quickly if it is stored for a long time under relatively high temperatures, moisture, contact with some metals, insufficient purification of fuel from fatty acids and glycerin);
- to reduce the pour point, the value of this characteristic for BD made of different raw materials is $(-2... +14)$ °C, while for diesel fuel according to the standards it is $(-25 ... -10)$ °C;
- chelating additives, since a prolonged contact of BD with such materials as copper, brass, bronze, lead, tin and zinc can lead to an increase in the rate of decomposition and consequently to the formation of deposits. Chelating additives linking heavy metals with organic compounds, which biodiesel fuel predominantly consists of, can reduce or even eliminate the negative effect of the presence of metals.

Wintron XC30 is an effective additive for reducing the BD pour point. Figure 4.9 illustrates the effectiveness of using Wintron XC30 as the additive for B100 made from rapeseed and deep frying oils. Before adding this additive, the fuel must be heated to the temperature of at least 5 °C above the cloud point. The required amount of Wintron XC30 is about .2–2% of the fuel mass; it depends on the type of raw material for B100. This additive is more effective for biodiesel fuels made from unsaturated oils, such as rape, sunflower, soybean ones. The use of Wintron XC30 is beneficial for frying fats, although their composition may vary considerably. Wintron XC30 is less effective for BD made of animal fats and palm oil [7].

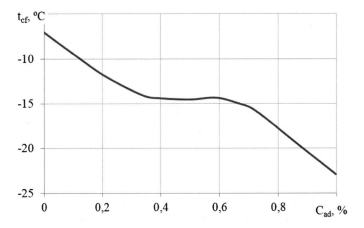

Fig. 4.9 Dependence of the temperature of b100 filter clogging from the content of wintron XC30 additive in fuel

The fuel pour point can also be reduced by cold filtration. The main disadvantage of this method is a decrease in the cetane number due to the removal of a part of molecules with a high cetane content. For the production of biodiesel fuels with low pour points, it is recommended to make fuel from raw materials with normal flow characteristics at low temperatures, for example, from rapeseed oil [7].

A number of safety rules are to be obeyed when using Wintron XC30. Since the additive contains toluene (2%), it is necessary to keep it as far as possible from potential sources of heat or ignition. It is necessary to avoid the contact of Wintron XC30 with the eyes, skin and clothing, respiratory tract. The additive should be stored in a closed container in the room with normal ventilation [7].

The Baynox® additive can be used to maintain the BD stability. It prevents the premature oxidation of unsaturated fatty acid esters and formation of volatile compounds and corrosive carboxylic acids in the fuel. Moreover, it prevents the early formation of polymerized precipitates which can lead to the damage of the engine parts and assemblies. The additive is a mixture of 20% of the highly purified 2.6-di-ter-4-methylphenol dissolved in the specially manufactured and purified rapeseed methyl ester that meets the quality standards DIN 51606 and EN 14214. The additive is not corrosive to metals and burns without the formation of residues. The recommended concentration is .1–1.0% of the fuel mass. Table 4.1 shows the main characteristics of the stabilizing additive Baynox® [7].

Fuel-clean BD is a biodiesel fuel additive on the enzymatic basis with a combined impact:

– biocidal effect: it quickly destroys microorganisms that cause decomposition of fuel and prevents their reproduction;
– fuel catalyst: it allows for a more complete combustion, reduction in emissions, improvement in fuel economy [8].

Table 4.1 Baynox® additive characteristics

Characteristic	Value
Physical state under normal conditions	Liquid
Content	1 L consists at least 200 g of the 2.6-di-ter-4-methylphenol dissolved in biodiesel fuel (min 225 g/kg)
Colour and smell	Clear pale yellow liquid with the BD smell
Formation of blends	Very well blendable with biodiesel fuel and organic solvents; does not mix with water
Density under 20°C	On average 0.89 kg/m^3
Viscosity under 20°C	9.9 mm^2/s

The required concentration in fuel is .01%. Fuel-clean BD is effective for both biodiesel and conventional diesel fuels. The additive does not cause damage to engine components, fuel lines and seals.

The use of Fuel-clean BD reduces the number of bacteria and fungi by 99.12%. Unlike other biocides, the use of this additive does not cause formation of chemical sediments and deposits. When it is applied, residual sediments dissolve and can then be burned in the engine without side effects. Besides, bacteria and fungi do not adapt to Fuel-clean BD.

Fuel-clean BD is very effective for blends of BD and DF. Application of the additive prevents delamination of the blend even during a prolonged storage.

Killem Biocide is another effective additive for elimination of microorganisms. It is basically a pesticidal mixture based on water, which is not used for fuel blends [8]. Killem Biocide destroys almost all bacteria and fungi that can multiply in BD. The National Renewable Energy Laboratory (USA) recommends using this additive at prolonged storage of biodiesel or when using new fuel tanks. Unlike other additives, Killem Biocide mixes exclusively with the aqueous phase of the fuel. 1 ml of the substance is sufficient for processing 20 L of BD. When the fuel is heavily contaminated with microorganisms, it is recommended to increase this amount by 2–3 times. Some characteristics of the additive are presented in Table 4.2.

The premises where tanks with the additives are stored should be well ventilated, with ventilation being artificial (general or local). It is recommended to wash the tanks thoroughly after storage of the substance. All spills and leaks of the substance are subject to mandatory elimination, which should employ special adsorbents.

The cloud point and the temperature of hardening of BD are higher than those of diesel fuel. According to the ASTM standard, no special requirements are imposed on the pour point of biodiesel fuels, but these data should be indicated by the fuel producers for potential customers.

The filter clogging temperature is an indication of the possibility of using BD during the cold season. This parameter is well correlated with the cloud point temperature. The pour point and cloud point are easy to measure, so they are usually used to characterize the possibility of using biodiesel at low temperatures.

Table 4.2 Killembiocide additive physical characteristics

Characteristic	Value
Boiling point, °C	Around 100
Water solubility	Full
Density under the temperature of 20°C, kg/m^3	1176
Colour and smell	Yellow and green with the smell of sulphur
Physical state under normal conditions	Liquid
Volatility standard	Same as that of water
Fire safety	If stored correctly, it does not self-ignite; at burning of a dry matter (without water adding), carbon, nitrogen and sulphur oxides may be released
Stability	It is a sufficiently stable substance under appropriate storage conditions
Hazardous polymerization	Does not occur
Incompatibility with other substances	Mineral acids: sulfuric, nitric, hydrochloride
Possible hazards during storage	Flammable gases are released during the substance oxidation; gaseous substances such as amines or hydrocarbon disulphide are released during its thermal decomposition

Various fuel additives can improve these parameters of BD and diesel fuel. The general effect of such additives is based on changing the shape and size of the wax crystals that slows crystal growth and reduces the pour point. The additives contain individual components, usually copolymers of ethylene and vinyl acetate or other olefin-ester copolymers. Fuel additives for BD and its blends which are currently present on the market reduce the pour point, i.e. the fuel can be used at low temperatures.

At the University of Idaho, the studies have been conducted to evaluate the effectiveness of various additives to reduce the pour point and cloud point in BD from soybean oil and its blends with summer diesel fuel [9]. Four additives of different manufacturers were used for the study: Flozol 503 (Lubrizol Corporation), Bioflow 875 (Octel Sterron), MCC P205 (MidContinental Chemical), Arctic Express .25% (Power Service). Testing of the effect of the additives involved the blends B5, B20 with summer diesel fuel (cloud point of −17 °C, pour point of −22 °C) and B100. The results of determination of the additives' influence on the cloud point and pour point temperatures of biodiesel fuel and its blends are shown in Fig. 4.10 [9].

After the growth of wax crystals slows down under the action of additives during the first use, they subsequently have the minimal effect on the cloud point (CP). The average decrease in CP is about 0...1.3 °C and practically does not differ for various additives (Fig. 4.10a).

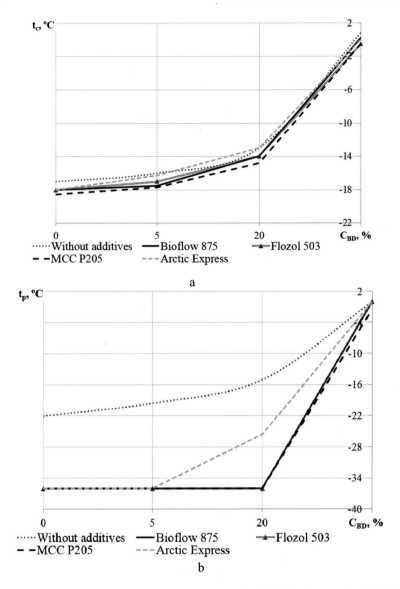

Fig. 4.10 Effect of Additives on the Temperature Properties of Biodiesel: **a**—on the cloud point; **b**—on the pour point

With the increase in the recommended amount of fuel additives, a small additional decrease in CP was observed twice; the highest value was recorded for the Arctic Express additive [9].

With the double amount of additives, the average decrease in the cloud point is 1.5 °C; it is about 1.0 °C with the recommended amount. With the tripled amount of additives, the CP decreased by 1.7 °C on average.

One of the most interesting results was the fact that the decrease in CP for B100 at 300% of the additive was the same as that at 200% of the additive or even lower. In general, when applying additives in the amount greater than recommended, the CP decreases for DF but increases for B100. Therefore, the double or triple amount of additives should be used only for blends with a low DF concentration. For B100, and blends with a high biodiesel content, this may have the opposite effect [9].

The tests of the additives' effect on the pour point (PP) were carried out until the temperature reached −36 °C. The PP was recorded at the temperature higher or equal to –36 °C, provided that the fuel did not acquire the gel-like structure (Fig. 4.10b).

The value of −36 °C was considered to be the lowest limit for measuring instruments in the research. Moreover, in most cases, the temperature is always above −36 °C under real operating conditions. When testing the additives, the pour point for diesel fuel and B5 blends was reduced to −36 °C or even lower. The same data were obtained for B20, with the exception of application of the Arctic Express additive. The effect of using the additives for DF, B5 and B20 with the aim of the PP reduction was not fully assessed; in fact, solidification of the fuels has not been achieved. However, the testing showed that the additives were practically not effective for the PP reduction in the pure BD, except for a slight decrease in the pour point when using the MCC P205 additive [9].

When the recommended amount of additives was doubled, the PP of B100 was reduced to −3 °C for all the additives except Flozol 503. At the addition of the latter, the pour point of B100 remained 0 °C. With the double amount of all additives, the PP of DF, B5 and B20 was −36 °C or even lower. When adding the triple amount of the recommended one, there was a gradual decrease in the PP when using only MCC P205 and Arctic Express. In these cases, the pour point of B100 decreased to −6 °C with the addition of MCC P205, and to −4.5 °C with the addition of Arctic Express. This indicates that these additives are more effective for DF than for soy BD. The observed decrease in PP for B5 and B20 was possible due to depression of diesel fuel. Although these additives were recommended for pure biodiesel, none of them affects B100 more efficiently than diesel fuel [9].

Fuel system materials. Since pure biodiesel fuel is a chemically and corrosively active liquid, it is incompatible with a number of coatings, adhesives, structural and sealing materials that are commonly used for diesel fuel systems. Thus, compounds made of lead, copper, bronze, zinc should be replaced with stainless steels and aluminum. It is undesirable to use natural rubber, nitrile and synthetic rubber as sealing materials [10].

When using B100 or blends with a high BD concentration (more than 50%), it is recommended to apply the equipment made of such materials as follows: austenitic stainless steels, polyethylene, polypropylene, riton (polypropylene sulphide, which is a high-quality glue used for equipment coming in contact with corrosive media), PVDF (vinylidene polyfluoride, a thermoplastic mass). The following substances can be used as sealing materials: nitriles, viton or synthetic rubber (fluoropolymer elastomer), PTFE or Teflon (polytetrafluoroethylene is one of the best antifriction materials), EPDM (ethylene-propylene rubber).

Tanks. The design of tanks for B100 does not differ from fuel tanks for diesel fuel. It is recommended to store fuel in aluminum and stainless steel tanks with nitride coating, fluorinated polyethylene and polypropylene, Teflon and fiberglass. Tanks for BD are manufactured by Aquastore and Southern Tank & Manufacturing Co; they are made of stainless steel [11, 12]. The vessels are made with account to the specific features of biodiesel fuel. Structural integrity of the tanks is very important because it is necessary to prevent water ingress, fuel leakage and multiplication of microorganisms. Tanks cannot be stored in the open under the sun, near sources of heat or possible ignition. Tanks can be supplied with insulation and a built-in heating system.

Pumps. Michael Smith Engineers Ltd. produces Finish Thompson pumps for pumping corrosive liquids including biodiesel. There are EP pumps (for low-viscosity liquids), PF and TBP pumps (for low and medium viscosity), TT (for medium viscosity) and HVDP (for high viscosity) pumps with electric (E) and pneumatic (P) drives. The main characteristics of the pumps are given in Table 4.3 [13]. The company Utah Biodiesel Supply specializes on the production of pumps as well.

Heaters. Among the equipment recommended for the BD heating, we can distinguish the Utah Biodiesel Supply and ArcticFox heaters. The former are basically flexible wide tapes which are "put on" the tank with fuel, i.e. external heating takes place (Fig. 4.11). The heater consists of a heating element located between two layers of fiberglass together with silicone rubber; it is equipped with an adjustable thermostat with the possible temperature range of (10...218) °C. The power capacity of the heating element is 1.2 kW. Its width is 102 mm, length is 1778 mm, and maximum temperature of the heating surface is 232 °C. [8]. Such heaters need not be built into the fuel system; they are very compact, corrosion-resistant, water-resistant and chemically resistant. They can be used on small vessels where BD is used.

The ArcticFox satellite heaters for the pipelines heating and submerged ones for fuel tanks are made of stainless steel. The liquid from engine cooling systems can be used as the coolant.

This manufacturer also produces pipelines with a built-in heating element that can be used as the fuel lines. Such an element with the power capacity of 150–600 W at the voltage of 12 and 24 V prevents the fuel from solidifying during transportation through pipelines. Self-adhesive electrical heaters are of interest, which can be attached to any metal surface. They are manufactured for the power capacity from 80 to 1000 W and the voltage of 12–220 V. Their aluminum heat exchange surface allows heating the fuel, ensuring its transfer from the gel-like to the liquid state, and maintaining it in this form [14].

Purification and mixing devices. The BD purification devices are produced by many companies, the most famous of which is Alfa Laval. A large lineup of separation and filtration equipment compatible with BD is produced by the German company Willibrord Lösing Filter-Technik. One of the products of this company is the separator SEPAR 2000 that have been approved and certified by such classification societies as Lloyds, Germanischer Lloyd, Bureau Veritas, and RINA. For the fuel separation

Table 4.3 Finish thompson pumps characteristics

Characteristics	Pump series				
	EP	PF	TBP	TT	HVDP
Pipe material	Polypropylene	Polypropylene, PVDF, stainless steel	Polypropylene, PVDF	Polyvinyl chloride, stainless steel	Stainless steel, viton, PTFE
Pipe diameter, mm	25	51	42	41 (polyvinyl chloride), 38 (stainless steel)	51
Head supply, l/min	23	150 (E); 83 (P)	75.7 (E); 47.3 (P)	26.5 (E); 60 (P)	26
Maximum viscosity of the pumped liquid, cPs	150	500	200	400 (E); 2000 (P)	100,000
Maximum density of pumped liquid, kg/m^3	1200	1800	1400	1800	1800
Maximum temperature of the pumped liquid, °C	49	49; 70; 105 (depending on the pipe material)	66	66	140
Pipe length, mm	69; 102	69; 102; 122	69; 102	69; 102; 122	69; 102; 122
Sealing material	Teflon with graphite filling	Teflon	PTFE with graphite filling	Teflon with graphite filling	–
Drives capacity, W	100	370; 640 (E); 370; 560 (P)	370	180 (E); 370; 560 (P)	–

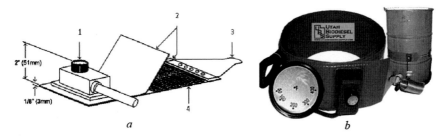

Fig. 4.11 Portable heater with external heating: **a**—arrangement; **b**—appearance separately and on the tank; 1—adjustable thermostat; 2—fiberglass reinforced with silicone rubber; 3—elastic clamping holder; 4—heating element

and filtration, there are produced combined units SEPAR 2000 (of the model SWK 2000) with the capacity of 5–130 l/min; some models are additionally equipped with the heating system that is automatically activated at low temperatures. Fiberglass is used as the material. The filter element can be used repeatedly during the service life of the unit. The installations are easily integrated into the fuel system. Figure 4.12 shows the hydraulic properties of the SWK 2000 models [15].

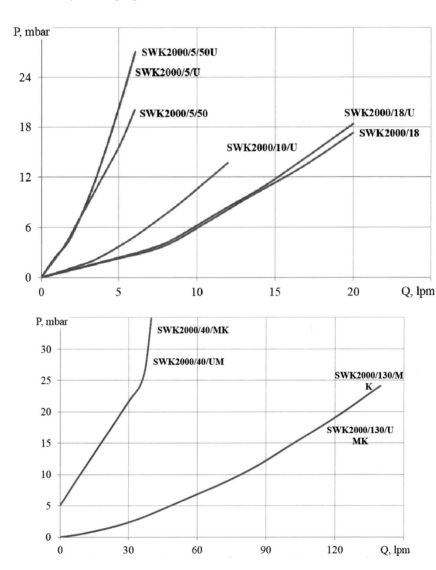

Fig. 4.12 Dependence of the pressure losses on fuel consumption of the SEPAR 2000 purification units

The SEPAR 2000 installations are well compatible with biodiesel fuels. The multi-stage centrifugal system makes it possible to purify fuel from water and mechanical impurities by almost 100%.

Separ Evo-10 is a fuel filter-separator constructed of polymer materials, which is easy enough to maintain. Here, fiberglass is used as the material; it can be subject to complete recycling with the possibility of subsequent use. The installation is relatively light and compact, which is important for use under ship conditions. The maximum throughput capacity is 600 l/h (10 l/min), the weight is about 1.1 kg, and the pressure losses are quite low. Figure 4.13 shows the principle of the installation operation and its design [15].

A substantial diameter of the filter cross section and channels allows minimizing the pressure losses and improving the efficiency of purification. The use of modern materials significantly reduces the weight and volume of the filter without a loss in efficiency or reliability. Due to the special design of the inlet and outlet nozzles, the unit is easily integrated into the fuel system. When the filter is used to prevent fuel contamination, the lid must be closed. Replacement and maintenance of the filter can be done manually.

The SEPAR 2000 purification equipment is certified by such maritime classification companies as Lloyds (No. 94/20036), Bureau Veritas (No. 1521 5842 A10 D), RINA (DIP /13/94) [15].

The use of compact combined treatment plants (filtration, separation, heating) and heaters as part of the BD fuel systems on ships reduces the amount of metal in the systems. It also increases their reliability by reducing the number of equipment pieces. Typically, technical documentation for pumps, filters, separators and other equipment indicates whether it is compatible with B100 ("Biodiesel compatible"). When using it under shipboard conditions, it is important to have quality certificates of the equipment issued by recognized maritime classification societies.

4.3 Parameters of Ship Biodiesel-Fueled Power Plants

When analyzing the possibility of the biofuel usage for water transport in general and in ship power plants in particular, the important aspect is a relatively small annual volume of such fuels produced in the world. While the reserves of natural gas are comparable to oil reserves, production of biodiesel fuel depends on the volume of secondary raw materials (oils and fats), the oilseeds are the primary raw materials for them. The possible volume of produced biofuels is directly related to the size of the acreage for oilseeds and the amount of oil that can be obtained from them. In this regard, it is of interest to analyze the potential volumes of biofuels produced.

According to the experts, edible oils remain the main raw material for the BD production. Nevertheless, the forecasts state that their share in total production will have decreased from 90% to 75% by 2019. This is due to the development of production of biodiesel fuels from jatropha (oil palm), which is not a food crop; its main

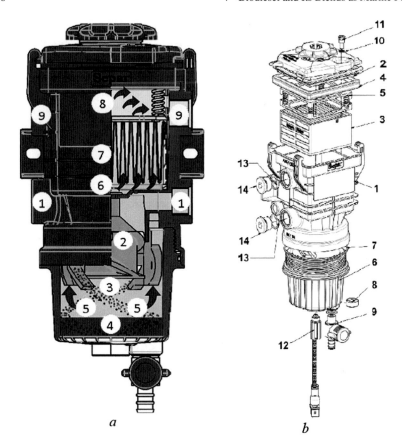

Fig. 4.13 Separ Evo-10 purification unit: **a**—operation principle: 1—fuel inlet; 2—intensive twisting of the fuel flow in the internal screw of the passive cyclone; 3—the fuel passes from the internal screw into the depositing tank; 4—due to the rotational energy, water and heavy solid particles are separated from the fuel and form a sediment at the bottom of the depositing tank; 5—the fuel is directed upwards into the pre-chamber of the filter; 6—increased cross section of the chamber reduces the speed of the fuel flow; 7—suspended particles and minute drops of water are retained on the folds by the filter medium of the filter element; 8—purified fuel enters the outlet chamber; 9—fuel outlet; **b**—installation design: 1—filter housing; 2—filter cover; 3—filter element; 4—cover gasket; 5—spring block; 6—depositing tank; 7—septic tank of the sediment bowl; 8—nut with two apertures; 9—drain valve; 10—USIT-washer; 11—depressurization bolt; 12—water level sensor; 13—O-washer; 14—cap

acreage is concentrated in India. Also, animal fats and biomass of the second generation (wood waste, biowaste, energy-intensive fast-growing plants, etc.) will be used more widely. The share of such raw materials by 2019 may reach 6.5% of the total volume (Fig. 4.14).

It is predicted that over 10 years the amount of BD produced will grow by more than 2.5 times compared to 2009 and will have reached 40 billion liters by 2019

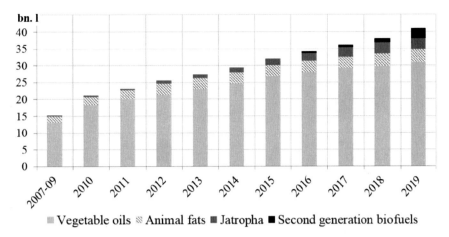

Fig. 4.14 Dynamics of world production of biodiesel fuels from various raw materials

[16]. Since biodiesel fuels are intended primarily for diesel engines of vehicles, the volume of their use on ships is likely to increase.

The results of the studies conducted by the US Department of Energy in 2008 suggested that in 2020 biofuel production in the world would be 54 billion gallons of ethanol equivalent (EGE), and in 2030—83 billion (Fig. 4.15).

The volume of production by 2020 will have grown by 4 times. The leading producers of biofuels in 2020 will be the United States and Brazil with more than half of the world production. The Western Europe will take the third place. It is projected that a significant part of biofuels by 2030 will have been produced in such

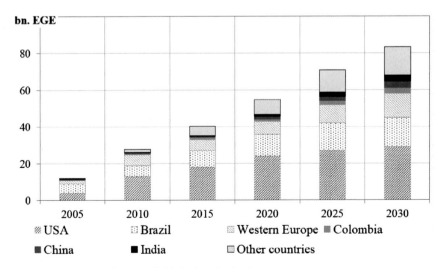

Fig. 4.15 Dynamics of changes in biofuel production by country

countries as China, India and Colombia. In the process of developing the model, a combination of relatively high oil prices with the availability of raw materials for production was adopted as a basic version shown in Figs. 4.15, 4.16 and 4.17 [17].

It is planned that the USA will not only be the leading producer, but also the largest consumer of biofuels, consuming more than half of the global level by 2020 (Fig. 4.16). Starting from 2005, the annual growth of 10% will have given a 10-fold increase in biofuel consumption in the USA by 2030. The second largest consumer

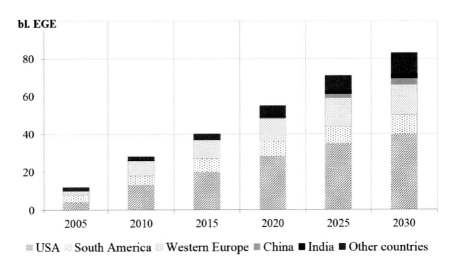

Fig. 4.16 Dynamics of changes in biofuel consumption by country

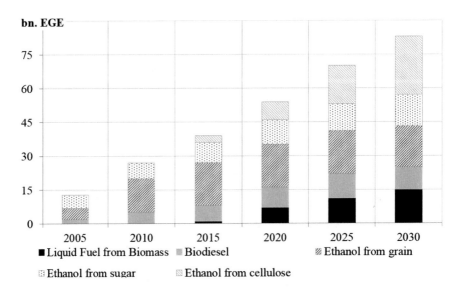

Fig. 4.17 Dynamics of changes in biofuel production from various raw materials

will be the Western Europe, followed by the countries of Central and South America. Brazil will remain the net exporter (i.e. the volume of biofuel exports will significantly exceed similar imports) [17].

In the structure of consumption, ethanol produced from grain and sugar will dominate until 2020. After those years, the share of biofuels from cellulose, both ethanol and other liquid fuels from biomass (LFB), will increase significantly (Fig. 4.17). More than 75% of ethanol from grain will be produced in the USA, while Brazil (about 80%) will retain the leading position in the production of ethanol from sugar [17].

An increasingly important role in meeting the world demand will be played by fuels from cellulose processed by biochemical transformations into ethanol or by the Fischer-Trop process into liquid hydrocarbons [17]. The share of cellulosic fuels will account for almost half of the world's biofuel production (49%). It is planned that more than a third of the production volume of such fuels will be accounted for by the USA; Brazil also has a considerable potential. BD will remain an important component of the global production of biofuels, especially in European countries, where the highest subsidies are set for it. By 2030, its share will have been 12%, which is less than that of ethanol from various raw materials and liquid hydrocarbons obtained by the Fisher-Trop process [17].

Further, the influence of biodiesel fuel and its blends directly on the ship power plant, including the fuel system, is considered in more detail.

Indicators of the SPP energy efficiency in the transition to biodiesel are not changing significantly; there may be a slight increase in specific fuel consumption and, as a consequence, the efficiency decrease. The use of blends with a low BD content practically does not affect the energy efficiency of the SPP.

When using BD on ships, there is no need to replace the engine or propulsion complex. The main mass-dimensional indicators that can be changed will be the mass and volume of fuel reserves and the mass of the SPP fuel system.

In accordance with some increase in the mass of the fuel reserves, the autonomy indicators when operating on biodiesel will be slightly lower than those for diesel fuel. This indicator will not be significant because of the short duration of the voyage during the operation of small vessels in the coastal and inland navigation areas.

The flash point of biodiesel fuel, depending on the standards, should not be lower than $100...130$ °C; depending on the raw material, it can reach $180...200$ °C. At the same time, the value of this characteristic for DF averages only at $58...64$ °C. DF and BD are classified as low-risk substances for the human body. The self-ignition temperature of DF is $230...300$ °C (in summer), of B100—$300...350$ °C. The presented data indicate that biodiesel fuel is safer than diesel fuel as for the flash point and the temperature of self-ignition.

Indicators of maneuverability can be estimated from the range of possible operating modes and characteristics of torque transmitted to the engine, as well as the fuel preparation rate.

During preparation of the fuels in the fuel system, the filtration and separation rates for BD are somewhat reduced in comparison with DF. As a result, the time of preparation for the engine start will increase. This can be avoided if the purification equipment installed in the FS is used directly for BD.

Economic indicators of the SPP when working on biodiesel will primarily be determined by the cost of this fuel, which depends on many factors, including primary raw materials and availability of subsidies for the BD production. The price of nuclear fuel is also subject to considerable fluctuations.

The BD use for the SPP leads to the improvement in its environmental performance due to the reduction in the level of emissions of virtually all the main components of exhausted gases but for nitrogen oxides [18].

The reliability of the engine and fuel system can be estimated by means of indirect indicators, such as lubricating properties of the fuel, interaction with various materials of the system and sealing. Figure 4.18 presents a comparison of the lubricating properties of DF, BD and their blends. Sample 1 corresponds to winter diesel fuel, and sample 2 corresponds to summer fuel. The test was carried out using the standard HFRR method (bench test for reciprocating motion) [19]. The obtained results indicate that the lubricating properties of B100 are much better than the ones of DF: there is almost twice less wear. Even a small addition of biodiesel to diesel fuel (.4 and 1%) significantly improves the lubricating properties of the latter.

There has been conducted a compatibility test of sealing materials (O-rings) in interaction with DF and B20 comparing to the contact of these rings with air [20]. The results of the test show that for various values of the destructive and shock loads, the overall parameters of the seals in contact with DF and B20 are practically the same [20].

Since biodiesel fuel in its physicochemical characteristics is close enough to diesel fuel, determination of the rational parameters of the SPP fuel system equipment when using this fuel includes the following tasks:

– to highlight the main characteristics of fuel systems to be changed;

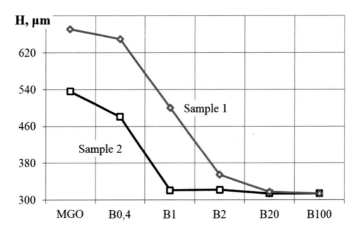

Fig. 4.18 Dependence of the level of the sample wear on the BD concentration

- to determine the change in the characteristics in case of using B100 from various raw materials, different B100 samples from one raw material, blends with a low BD content (up to 20%), blends with the BD content of 20 to 100%;
- on the basis of the analysis of the data obtained, to provide recommendations regarding determination of the characteristics of universal and bi-fuel systems.

During the research, the following characteristics of the fuel system equipment were determined: the mass and volume of fuel reserves for the voyage, power capacity of the fuel pumps drives, filtration and separation rate. The values of the system characteristics are compared with the use of B100 from different raw materials, pure biodiesel from the same raw material of different samples, blends with a low BD content (up to 20%) and the BD content of 20–100%. The calculations were performed according to the materials presented in [21–23].

Figure 4.19 represents the results for the following fuels: MGO (marine gas oil), fatty acid methyl esters (FAME), rapeseed methyl ester (RME), soybean methyl ester (SME), palm oil methyl esters (POME), sunflower oil methyl esters (SOME), tallow methyl esters (TME). Physical and chemical characteristics of biodiesel fuels (fatty acid methyl esters) were taken in accordance with the data presented in [24] for FAME, SME, POME and SOME, in [25] for TME and RME.

The data are given in the relative form; the corresponding values for DF have been taken for the unit of comparison. The analysis of the obtained characteristics indicates that, depending on the raw material, the required fuel mass can increase by almost 17%, the required relative volume of fuel tanks—by 10%, which is associated with the lower heat value of biodiesel fuels. For pumping of B100, the required power capacity of the pumps can be increased up to 50%. The filtration rate can be reduced by 2.5 times, while the separation rate can be reduced by almost three times [2].

The change in the relative characteristics of the fuel systems using fuels with the BD content from 0 (pure DF) to 100% (B100) is shown in Fig. 4.20. The indicators were determined using the low-sulphur diesel fuel and biodiesel from soybean for two samples. The first sample was DF with the cetane number of 47.1 and B100 with the cetane number of 55, the second sample those with 43.5 and 51.1, respectively. The percentage of BD in the blend is determined on the axis of abscissas. More detailed characteristics of these fuels and their blends are given in the paper [26].

The data shown in Fig. 4.20 indicate that the characteristics of the fuel system vary proportionally depending on the concentration of biodiesel fuel in the blend and have a practically linear dependency. The relative mass of the fuel reserves increases by almost 16%, the volume of fuel tanks—by 10%, the power capacity of the pump drives—by almost 50%. The filtration and separation rates decrease by 35...40% and 50...55%, respectively.

The FS parameters have been compared for the use of B100 of 27 different samples from one raw material. The physicochemical characteristics of the samples meet the international quality standards [27]. The difference in the density values of the considered fuel is up to 1.5%, those of viscosity—up to 18.2%. The capacity of the fuel tanks and filtration rate change proportionally to the density and viscosity, by 1.54% and 18.4%, respectively. The power capacity of the drives of transfer

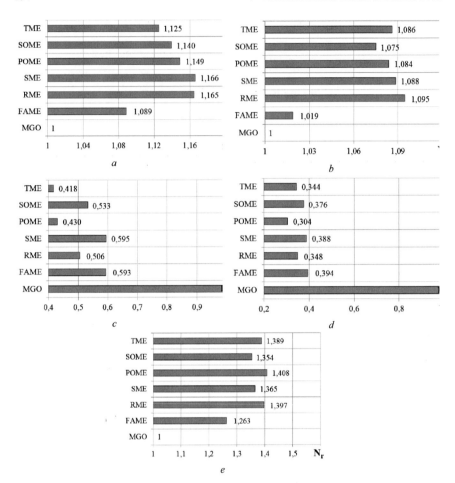

Fig. 4.19 Characteristics of the SPP fuel systems when using diesel fuel and biodiesel from various raw materials: **a**—relative mass of fuel reserves; **b**—relative capacity of fuel tanks; **c**—relative filtration rate; **d**—relative separation rate; **e**—relative capacity of the transfer pump drive

pumps for different types of B100 increases by only 3.5...4%, while the filtration and separation rates are almost 22 and 29%, respectively. The results of calculations in the relative form, as well as the density and viscosity of the samples, are shown in Fig. 4.21 [2].

Similar calculations have been carried out for the blends of diesel and biodiesel fuel with the BD content of up to 30%. Such blends are most often used in engines, and their physicochemical characteristics are presented in the paper [27]. Most samples are B20. Depending on the content of biodiesel fuel in the blend (7–28%) and its properties, the mass of the fuel reserves increases by 2.7% compared with diesel fuel, the capacity of the fuel tanks increases by 1.7%, the power capacity of the

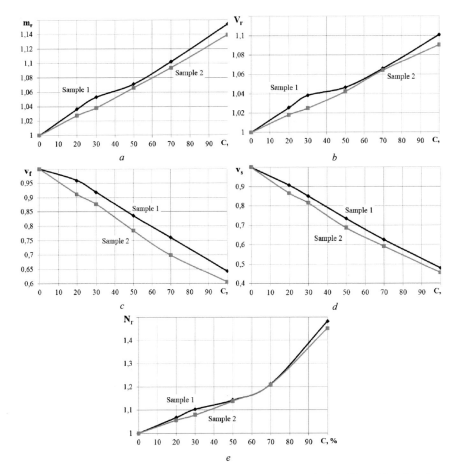

Fig. 4.20 Characteristics of the SPP fuel systems when using diesel fuel, B100 and Their Blends in Different Proportions: **a**—relative mass of the fuel reserves; **b**—relative capacity of the fuel tanks; **c**—relative filtration rate; **d**—relative separation rate; **e**—relative capacity of the transfer pump drive

pump drives increases by 8.7%, and the filtration and separation rates are reduced by 22.3% and 24%, respectively (Fig. 4.22).

4.4 Conclusions

1. There have been developed variations of the diagrams of fuel systems in terms of BD application. In one of the variations, the universal FS for DF, BD and their blends in different ratios is provided on the vessel on condition that the fuels are used on the vessel alternately, and ready DF and BD blends are received, and

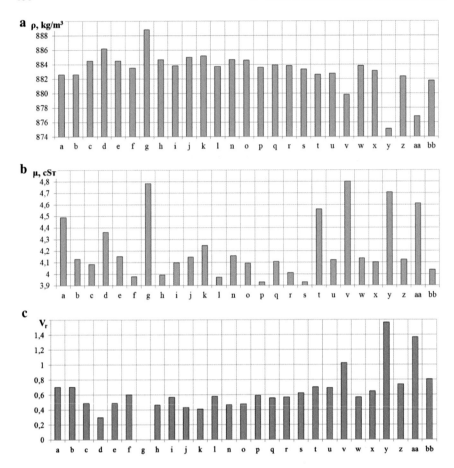

Fig. 4.21 Characteristics of the SPP fuel systems when using different samples of B100 (for the sample with the lowest value of the characteristic it was assumed to be 0, for other samples the values were compared with the lowest one): **a**—density, kg/m³; **b**—viscosity, cSt; **c**—relative capacity of fuel tanks; **d**—relative filtration rate; **e**—relative separation rate; **f**—relative power capacity of the transfer pump drive

not prepared onboard. It is most suitable for small vessels with rigid mass-size restrictions. In the second variation, by analogy with the use of heavy and light oil fuels, it is suggested to use two fuel systems with mutual reservation of individual equipment and preparation of fuel blends directly on board of the ship.

2. There have been developed recommendations regarding the features of mechanical and chemical preparation of biodiesel fuels on ships, as well as preparation of the BD and DF blends under ship conditions. It is shown that the equipment chosen for biodiesel fuel systems can be the same as that for diesel fuel systems; the main condition is the use of materials compatible with BD.

3. The main raw materials for the BD production have been and still are edible oils. Nevertheless, their share in total production will have decreased from 90%

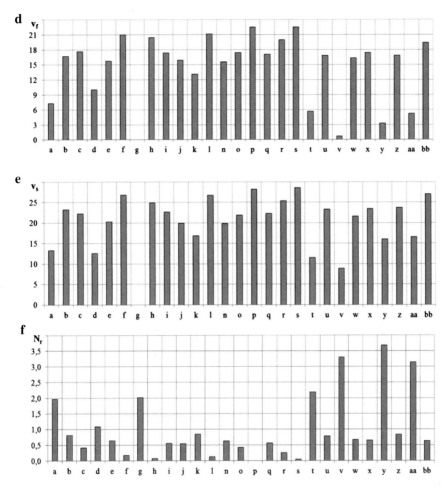

Fig. 4.21 (continued)

to 75% by 2019. Partial replacement of edible oils with jatropha, algae, and biomass of the second generation is projected. The share of such substitute raw materials by 2019 may reach 6.5% of the total volume. It is forecasted that by 2019 the amount of produced BD will have been 40 billion liters.

4. The range of rational parameters of the SPP fuel systems for biodiesel fuels is determined during the design and selection of equipment. It has been established that the use of biodiesel fuels, compared to diesel, leads to the need in the increase

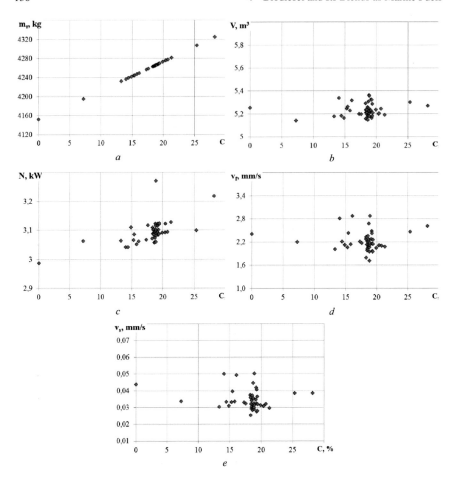

Fig. 4.22 Characteristics of the SPP fuel systems when using diesel fuel and its blends with BD content of up to 30%: **a**—mass of fuel reserves, kg; **b**—fuel tanks capacity, m³; **c**—power capacity of the transfer pump drive, kW; **d**—filtration rate, m/s; **e**—separation rate, m/s

of the mass of fuel reserves up to 16%, volume of the fuel tanks—up to 10%, power capacity of the pump drives—up to 50%, while the filtration and separation rates are reduced by 60 and 65% respectively. For blends with a low BD content (up to 20%), the capacity of fuel tanks and the mass of fuel reserves increase insignificantly (by 1.7 and 2.7%), the energy consumption for the fuel pumping increases by 8.7%, and the filtration and separation rates are reduced by 22 and 24%.

References

1. R. Sadler (2008) Biofuels and their effect on the shipping industry [Online]. Available: https://euroshore.com/sites/euroshore.com/files/documents/biofuels%20in%20shipping.pdf.
2. V. M. Gorbov, V. S. Mitenkova, "Obgruntuvannya ratsional'nykh parametriv palyvnykh system SEU pry vykorystanni biodyzel'nykh palyv [The substantiation of SPP fuel systems rational parameters with biodiesel fuels utilizing]", Visnyk Natsional'noho tekhnichnoho universytetu "Kharkivs'kyy politekhnichnyy instytut", Tematychnyy vypusk: "Enerhetychni ta teplotekhnichni protsesy y ustatkuvannya", pp. 180–186, #3, 2009.
3. V. S. Mitenkova, "Efficiency Increasing of Alternative Fuels Utilization in Ship Power Plants", Ph.D. dissertation, Dept. Engines and Power Plants, National University of Shipbuilding, Mykolaiv, Ukraine, 2010.
4. V. M. Gorbov and V. S. Mitenkova, "Systema pidhotovky dyzel'noho ta biodyzel'noho palyv dlya dvyhuniv vnutrishn'oho zhoryannya [System for treatment of diesel fuel and biodiesel for internal combustion engine] ", Ukraine Patent 51960, Aug. 10, 2010.
5. A. Langevin, BioMer Project biodiesel for ships. Toronto, 2005, 21 p.
6. V. M. Gorbov, V. S. Mitenkova, "Sostoyanie i perspektivy ispol'zovaniya biodizel'nyh topliv v sudovoj ehnergetike [Situation and perspectives of biofuels using in marine engineering]", Vestnik SevGTU: Mekhanika, ehnergetika, ehkologiya, pp. 107–112, Is. 97, 2009.
7. Wintron® XC30 [Online]. Available: http://www.biofuelsystems.com/other/wintronxc30.pdf.
8. Utah Biodiesel Supply [Online]. Available: www.utahbiodieselsupply.com.
9. Impact of additives on cold flow properties of biodiesel [Online]. Available: http://www.uiweb.uidaho.edu/bioenergy/ NewsReleases/TechNote3.pdf.
10. R. von Wedel (1999) Technical handbook for marine biodiesel in recreational boats [Online]. Available: http://www.cytoculture.com/Biodiesel% 20Handbook.htm.
11. Tanks for BioEnergy storage [Online]. Available: http://cstindustries.com/products/bioenergy-tanks/#page=page-1.
12. Biodiesel tanks [Online]. Available: http://southerntank.net/biodiesel-tanks.htm.
13. Michael Smith Engineers product selector [Online]. Available: http://www.michael-smith-engineers.co.uk/ pumps.htm.
14. Heating solutions for biodiesel applications [Online]. Available: http://www.arctic-fox.com/sitepages/pid74.php.
15. Fuel filtration and fuel polishing systems [Online]. Available: www.separ.co.uk.
16. Biofuel production 2010–2019 [Online]. Available: http://www.oecd.org/ document/9/0,3746,en_36774715_36775671_45438665_1_1_1_1,00.html.
17. World biofuels production potential. Understanding the challenges to meeting the U.S. renewable fuel standard [Online]. Available: http://www.osti.gov/bridge/servlets/purl/946833-Glee7C/946833.pdf.
18. BioMer: biodiesel demonstration and assessment for tour boats in the Old Port of Montreal and Lacine canal national historic site [Online]. Available: http://www.sinenomine.ca/Download/BioMer_ang.pdf.
19. Lubricity benefits [Online]. Available: http://www.biodiesel.org/pdf_files/fuelfactsheets/Lubricity.PDF.
20. E. Frame, R. L. McCormick (2005) Elastomer compatibility testing of renewable diesel fuels [Online]. Available: http://www.nrel.gov/vehiclesandfuels/ npbf/pdfs/38834.pdf.
21. [21] G. A. Artemov, V. P. Voloshin, A. Ya. Shkvar, V. P. Shostak, Sistemyi sudovyih energeticheskih ustanovok [Ship power plant systems]. Leningrad: Sudostroenie Publ., 1990, 376 p.
22. [22] V. V. Voznitskiy, Praktika ispolzovaniya morskih topliv na sudah [Practice of marine fuels using on ships]. Sankt-Peterburg: Biblioteka sudovogo mehanika Publ., 2006, 124 p.
23. YU. I. Dytnerskij, Processy i apparaty himicheskoj tekhnologii, Part 1. Teoreticheskie osnovy processov himicheskoj tekhnologii. Gidromekhanicheskie i teplovye processy i apparaty [Process and equipment of chemical engineering. Part 1. Theoretical basis of chemical engineering process. Hydromechanical and thermal processes and equipment]", Moskva: Chemistry Publ., 1995, 400 p.

24. D. L. Clements (1996) Blending rules for formulating biodiesel fuel [Online]. Available: http://
    biodiesel.org/resources/reportsdatabase/reports/gen/19960101_gen-277.pdf.
25. S. Lebedevas, A. Vaicekauskas, "Research into the application of biodiesel in the transport
    sector of Lithuania", TRANSPORT, pp. 80–87, V. XXI, # 2, 2006.
26. The physical & chemical characterization of biodiesel low sulfur diesel fuel blends [Online].
    Available: http://www.biodiesel.org/resources/reportsdatabase/ reports/gen/19951230_GEN-
    253.pdf.
27. McCormick R. L., Alleman T. L., Ratcliff M. and others (2005) Survey of the quality and
    stability of biodiesel and biodiesel blends in the united states in 2004 [Online]. Available:
    http://www.nrel.gov/docs/fy06osti/38836.pdf.

Chapter 5
Synthetic Coal-Based Fuels and Their Combustion

5.1 Integrated Plasma Coal Gasification Power Plant

Coal as a feedstock for power generation becomes more and more attractive in case of its preliminary conversion into a gaseous fuel named synthesis gas (SG). Plasma assisted coal gasification appears to be the best solution for portable and small to medium-scale processing facilities [1–7].

A combined cycle power plant with integrated coal gasification (IGCC) promises higher efficiency when compared to conventional, steam-cycle coal-fired power plants. As a further development of the IGCC, we offer the so-called Integrated Plasma Gasification Combined Cycle (IPGCC). This technology was offered and described in detail in paper [8]. It seems appropriate now to consider the efficiency of a power plant with high-ash coal plasma gasification and further burning of the obtained synthesis gas in a gas turbine engine with power production and deep heat utilization.

A general schematic of a technological complex for high-ash coal processing into SG with further power production is developed (Fig. 5.1) [9].

It enables the study and comparison of power plant efficiency with three methods of coal gasification in a direct-flow gasifier with plasma process initiation: (1) air gasification, (2) oxygen gasification, and (3) water steam + oxygen gasification. These three oxidizers were selected based on the performed analyses reported in paper [10].

The proposed plasma coal technology has the following innovative features [8, 9].

1. Employment of a recently-developed new generation of high power plasma torches with virtually unlimited lifetime due to electrodeless design, high efficiency of plasma production (up to 80-86%) due to application of the solid state power supplies currently under construction, and utilization of the patented reverse vortex plasma stabilization, which allows 90–95% torch thermal efficiency [11].

© Shanghai Scientific and Technical Publishers 2021
X. Yang et al., *Alternative Fuels in Ship Power Plants*,
https://doi.org/10.1007/978-981-33-4850-9_5

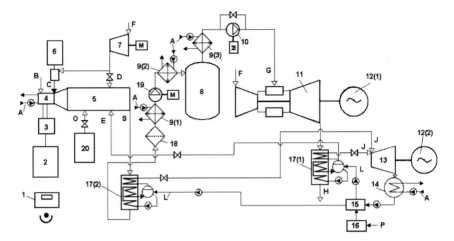

Fig. 5.1 Schematic flow diagram of a combined cycle power plant with integrated plasma coal gasification. Units: 1—control system; 2—plate power module of the RF power source; 3—RF module; 4—RF plasma torch; 5—multi-stage coal gasifier; 6—coal mill and feeder; 7—air compressor for coal transportation; 8—synthesis gas storage tank; 9 (1), 9 (2), 9 (3)—synthesis gas cooler; 10—fuel synthesis gas compressor; 11—gas turbine engine (GTE); 12 (1,2)—power generators; 13—steam turbine (ST); 14—steam condenser; 15—hot well; 16—water treatment system; 17 (1,2)—heat-recovery steam generators (HRSG); 18—synthesis gas treatment module; 19—synthesis gas compressor; 20—oxygen production module. Working media: A—cooling water, B—plasma gas; C—coal dust; D—air for gasification; E—water steam for gasification; F—atmospheric air; G—fuel synthesis gas; H—GTE exhaust; J—overheated water steam; L—feed water HRSG; O—purging oxygen; P—fresh water; S—synthesis gas after the gasifier

2. Plasma coal gasification provides simple and reliable process initiation due to high temperature and the reactivity of the plasma plume, dramatic reduction in processing time, corresponding reductions in gasifier volume and weight, and the ability to use water steam and oxygen as the oxidants.
3. Presently-developed plasma torches can be scaled up from output power levels of 500 kW—10 MW per unit, and use a wide variety of plasma-sustaining gases [11].
4. Elimination of supplemental fuel requirements.
5. High degree of thermal energy utilization in a heat utilization loop.
6. Potential for additional thermal efficiency improvements, for example, by using a synthesis gas afterburner.
7. Opportunity to install such a unit near coal mines to consume the coal locally, thereby avoiding significant transportation and environmental costs.

Multistage coal gasifier startup and operation are provided by the high-frequency plasma torch 4, which generates a plasma jet with an average temperature of about 4500 K. Depending on the gasification method, air, oxygen, or steam can be used as the plasma feedstock gas. Operation of the plasma torch is provided by a high-voltage power supply source 2 with a high-frequency module 3.

When using the air gasification method, purge air is supplied by the air compressor 7 with the ratio of 28:1 in relation to the plasma feedstock air flow rate, and the ratio of 2.8:1 in relation to coal consumption. The synthesis gas mass output ratio is 3.65:1 in relation to the consumed coal. If the oxygen or oxygen-steam gasification methods are considered, the oxygen production module 20 and its supply system must be added to the flow diagram.

Large quantities of oxygen should be produced using the oxygen gasification technology. Its mass fraction is about 45% of the gasified coal flow. For oxygen-steam gasification, mass flow of the oxygen supplied into the oxygen gasifier is considerably smaller (34–38%). The ratio of SG output to coal input mass flow is approximately 1.4:1.

The described ratios of mass flows of the process air, plasma air, oxygen and synthesis gas in relation to the coal consumptionare achieved after preliminary variant calculations of the gasification products content using the TERRA code [12], with definitions of gasifier's operational parameters to achieve maximum process efficiency.

During the process of coal oxygen + steam gasification, water steam is additionally supplied into the gasifier. This steam is generated in the heat-recovery steam generator (HRSG) 17(1) of a heat-utilizing loop (HUL) of the gas turbine engine (GTE) (11). Superheated steam with the temperature of 550–650 K can be fed in the mass quantity of 15–25% depending on the volume of coalconsumption. For this scenario the SG to coal mass output ratio is approximately 1.5:1.

Synthesis gas which was produced in the gasifier 5 is supplied into the HRSG 17 (2), where it cools down to 380-400 K to produce superheated water steam with the same parameters as in the HRSG 17(1). Then it passes the synthesis gas purification module and the synthesis gas cooler 9 (1), where its temperature drops to 290–300 K. The obtained SG is then compressed to a pressure of 1.4–1.6 MPa by the compressor 19. It then goes through the second gas cooler 9 (2) to the synthesis gas storage tank 8.

Synthesis gas is supplied out of the tank 8 through the cooler 9 (3) directly or via a fuel SG compressor into the GTE 11, which generates power by means of the power generator 12 (1) in a quantity sufficient to sustain the production cycle and with excess sufficient for delivery to external consumers. Power produced by the GTE can be delivered to consumers (electric generator, plasma torch, SG compressor, pumps of the heat recovery system, etc.). This is net thermal power with respect to the coal.

After the GTE 11, the SG combustion products are used in the HRSG 17 (1) of the main HUL. Superheated steam generated in this HRSG is mixed with steam from the HRSG 17 (2), and then supplied to the heat recovery steam turbine 13, which drives a separate power generator 12 (2).

The steam turbine operates using overexpansion, discharging the waste steam into the vacuum condenser 14, in which the water loops of both HULs are closed. Condensate drains into the hot well 15, which feeds both HRSG 17 (1) and 17 (2) with water, as needed. Makeup water for the technological cycle is provided by the water treatment system 16.

To conduct research and choose the appropriate design parameters, we have developed a mathematical model and calculation software. This approach allows us to simulate varying conditions including the quality of input coal fuel, the gas used to produce the plasma, and equipment selections, such as the gas turbine and other components. To determine the optimal coal to oxidant ratio for the plasma gasification process and composition balance of the thermo-dynamic systems for all of the above-mentioned gasification methods, and in order to produce SG with the maximum calorific value, the computer code TERRA was used [12]. This software allows us to define the volume fractions of the components and temperature of the gasification process with a given ratio of raw coal and an oxidant. TERRA has been successfully used for prediction of thermodynamic parameters in different energetic systems. Validation of this model is provided in [13], where the processes of plasma gasification of coal and petrocoke were studied.

In the following simulation, Taldykolsky coal from Kazakhstan (ash content 24%, moisture 18.0%, higher calorific value 19119.81 kJ/kg, air stoichiometric value 7.235 kg/kg) was considered as the feedstock.

Figure 5.2 shows the results of oxygen-steam gasification of this coal at the mass ratio $H_2O/O_2 = 25\%$.

The calculation software module for SG physical properties allows calculation of all thermo-physical properties required for further investigations: lower calorific value and properties of synthesis gas combustion products. Selection of the optimal coal to oxidant mass flow ratio correlation in the multistage direct-flow gasifier with plasma process initiation is performed using theoretical gasification efficiency defined through the equation:

$$\eta_g = \frac{m_{sg} H_{sg}}{H_{coal}}$$

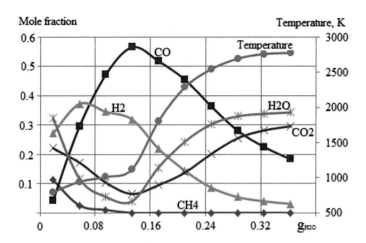

Fig. 5.2 Synthesis gas composition and temperature at the mass ratio $H_2O/O_2 = 25\%$ in the function of the oxidant weight content (g_{H2O})

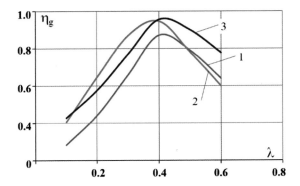

Fig. 5.3 Dependences of theoretical gasification efficiency on the oxidizer excess coefficient λ: 1—air gasification; 2—oxygen gasification; 3—steam-oxygen gasification with $H_2O/O_2 = 56\%$

where m_{sg} is amount of mass fractions of synthesis gas combustible components, H_{sg} is the lower calorific value of synthesis gas produced, H_{coal} is the lower calorific value of coal.

The results of calculations for various gasification technologies are given in Fig. 5.3. These results help define the optimal values of the oxidizer excess coefficients: 0.41 for air gasification, 0.39 for oxygen gasification, 0.42 for the oxygen-steam gasification with specified steam to oxidant ratio.

Calculation of the GTE parameters are performed for a specific type of a GTE, which is planned to be used in the technological process. After verification of the GTE mathematical model using manufacturer's specifications, the SG flow rate was determined for a nominal operation mode of the GTE compressor with the same level of gas temperature at the turbine inlet (rotor inlet temperature T_3). The parameters of the GTE operation and its capacity were determined; adjustments to the turbine's first stage nozzle to provide flow capacity have been estimated. Electric power generated by the GTE is also defined.

Calculations are performed to obtain SG production values, optimal parameters of the steam turbine heat-utilization loop, optimal parameters of water vapor pressures and superheating temperatures, and reaction self-sustainment ST and energy consumption requirements. Ancillary power needs and consumable usage rates are also modeled.

The parameters of the HUL, through which the SG exits the plasma gasifier at the temperature of 1000-1150 K, are calculated. Electric power generated by the heat recovery ST from HRSG superheated steam 17 (2) and internal consumption needs of the HUL (Fig. 5.1) are defined.

Power consumption for the compressor drive (7) that supplies air into the gasifier is defined. Power expenditure for the compressor (19) drive which pumps cooled SG into the storage tank (8) is also defined, as well as power consumption to drive the fuel SG compressor to feed it into the GTE. We determined consumption values for cases where SG pressure in the tank (8) dropped below the value in the GTE's combustion chamber. A centrifugal compressor with an electric motor drive was initially selected for this purpose.

Specific power consumption for 1 kg of oxygen was defined by totaling values for its separation provided by Air Products and Chemicals, Inc., compression in a separate compressor, and feeding into the gasifier. Similarly, we estimated the specific power consumption for 1 ton of pulverizing coal, including the coal mill (6) operation.

Operating efficiency of the entire complex on high-ash coal processing into SG is estimated through the efficiency of initial coal heat energy conversion into marketable (net) electric power

$$\eta_e = \frac{N_e}{H_{coal} G_{coal}} \tag{5.1}$$

where G_{coal} is the coal consumption.

Effective power output N_e in formula (5.1) is the amount of electricpower which can be sold, and is defined by the formula

$$N_e = N_{GT} + N_{ST} - N_{tech} \tag{5.2}$$

where N_{GT} is the electric power received from the GTE, N_{ST} is the electric power generated by the heat-recovery ST.

Total electric power consumption in the technological process N_{tech} in formula (5.2) is defined as the total power needs of the compressor, plasma torches, all pumps which enable operation of the main HUL (after gas turbine) and the HUL which is located after the gasifier, oxygen production and feeding, and coal grinding.

Relative power consumption in the technological process $\overline{N}_{tech}$ shall be defined through the formula

$$\overline{N}_{tech} = \frac{N_{tech}}{N_{GT} + N_{ST}}.$$

We have calculated the performance parameters of the plant, defined all of the required process parameters for the technological equipment, and specific power consumption for three different simple cycle GTE developed by the Gas Turbine Research & Production Complex "Zorya"-"Mashproekt", Ukraine [14]: UGT2500 (nominal power 2.85 MW, maximal cycle temperature 1224 K, efficiency 28.5%), UGT3200 (3.4 MW, 1273 K, 31.5%), and UGT3200RG (3.4 MW, 1273 K, 40%).

Effective GTE power output ranges from 2740 kW to 3720 kW for a power plant with the UGT2500 engine. A heat-recovery steam turbine produces near 30% of all power; and the efficiency of initial coal heat energy conversion into marketable power is 26.4–29.4%.

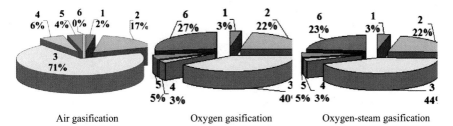

Fig. 5.4 Structure of the energy balance for a power plant with the UGT3200 gas turbine: 1—energy consumption of the heat recovery steam generators 1 and 2; 2—energy consumption of the plasma torches; 3—SG compressor power; 4—air blower power; 5—energy for coal grinding; 6—energy for oxygen production and feeding

The balance of power consumption for the own needs of the complex of high-ash coal processing into SG is shown in Fig. 5.4. The calculations have been done for the GTE of the UGT3200 type with the nominal power capacity of 3400 kW.

The calculations were performed for the direct-flow gasifier of coal with plasma process, initiating under the pressure of 0.15 MPa inside the unit, so the highest required power consumption was for the SG compressor drive—40–71%.

Electric power consumption for oxygen production and supply for two schemes with oxygen purging is 23–27%, while the power needs for the plasma torches operation is 17–22%. The overall power consumption for internal needs is 25–32%.

Effective power output is correlated with the power of the original GTE and ranges from 3250 kW to 4400 kW. A heat-recovery steam turbine produces from 20% to 30% of all power, and the efficiency of initial coal heat energy conversion into marketable power is 28–31%.

We have also considered the effect of pressure levels in the gasifier on the power plant efficiency. The efficiency parameters and power consumption values for the power plant needs were calculated for a pressure level up to 1.05 MPa. In Fig. 5.5, the results of this research are provided for the process of oxygen-steam gasification using the GTE UGT3200.

For the simple cycle gas turbine engine model, the optimal pressure level in the gasifier, which provides maximal efficiency of coal into electric power conversion, is about 0.8 MPa.

To define how the efficiency of the GTE affects the efficiency parameters of the power plant, additional investigations were conducted for a more efficient GTE (Table 5.1).

There was considered the UGT3200RG engine, which has a more complicated thermodynamic cycle (with regeneration of its exhaust gases its efficiency is 40% while its power output is 3,400 kW under ISO conditions). In Table 5.1, the balance of power produced by the power plant and the efficiency of initial coal heat energy conversion are shown for the direct-flow gasifier under the pressure of 0.15 MPa.

We have found that at the same power capacity of the original GTE, the power generated by the plant remains approximately at the same level and ranges from

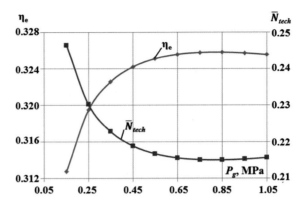

Fig. 5.5 Dependence of efficiency of the conversion of initial coal heat energy into the net electric power (η_e) and relative power consumption of the technological process ($\bar{N}_{tech}$) on pressure in the gasifier (P_g)

Table 5.1 Power plant characteristics

Gas turbine type	UGT3200RG		
Gasification media	Air	Oxygen	Oxygen-steam
Gas + turbine power output, kW	3856.0	3138.0	3142.0
Power output from a heat recovery steam generator, kW	1235.14	815.45	800.42
Total power output, kW	5091.1	3953.5	3942.4
Power consumption for heat recovery steam generator 1 operation, kW	24.19	18.64	17.74
Power consumption for heat recovery steam generator 2 operation, kW	9.40	4.29	4.75
Power consumption for plasma torches, kW	273.33	211.29	200.79
Power of a synthesis gas compressor, kW	741.05	256.67	258.30
Power of air blower, kW	100.60	26.87	25.54
Power for coal grinding, kW	60.13	46.48	44.17
Power for oxygen production and feeding, kW	0.0	260.3	209.8
Total energy balance for the plant needs, kW	1208.71	824.57	761.11
Coal consumption, kg/h	2405.4	1859.3	1766.8
Power available from coal, kW	11,898.2	9197.2	8739.8
Electrical power output, kW	3882.4	3128.9	3181.3
Power plant net thermal efficiency	0.3263	0.3402	0.3640

3140 to 3860 kW. The efficiency of initial coal heat energy conversion increases and is 32.6-36.4%. Significant reduction of relative power consumption for the complex needs was also observed, which is 19-24% of total power generation.

Decrease of the coal feed rate by a magnitude of about 1.36 was observed when switching from air to oxygen-steam gasification. This may be explained by a higher content of hydrogen and, correspondingly, a higher calorific value of the synthesis gas.

Based on the data presented, we can formulate the following conclusions.

1. The total (net) thermal efficiency of coal conversion into net electric power using direct flow gasification with plasma initiation and a low-power GTE with a relatively low rotor inlet temperature $T_3 = 950$–1000 °C was in the range of 25–36%.
2. The net electric power generation is in the range of 90–125% of the GTE ISO power capacity. Heat recovery contributes 20–30% of the total power.
3. The best results were obtained with the SG produced in a plasma gasifier with a mixture of steam and oxygen as the oxidants.
4. The influence of the pressure level in the gasifier on the efficiency and parameters of the entire complex have been studied. Elevated pressure levels in the gasifier (up to 0.75-0.8 MPa) could be considered as optimal for the developed IPGCC scheme with the considered simple cycle gas turbines.
5. Despite a relatively low power plant thermal efficiency, which is defined, first of all, by much lower efficiency of low power gas turbines in comparison to high power ones, our technological approach provides the following advantages: clean and economically feasible conversion of solid fuel into combustible gas, portability (could be placed in standard sea containers), readiness to commercialization due to implementation of critical components already in production, and a relatively low cost of ownership.
6. For higher electrical power output systems, more efficient simple cycle gas turbines with the rotor inlet temperature $T_3 \geq 1250$ °C, and without CO_2 capture, the net thermal efficiency will be much over 40% for oxygen-steam plasma gasification.
7. The main directions of internal power consumption reduction in the IPGCC power plant are: (1) utilizing more efficient GTE and ST, (2) increase in the gasification process pressure (that allows for significantly decreased power requirements of the synthesis gas compressor), (3) development of plasma systems with the total plasma production efficiency of about 80–85%, mainly by application of the solid state power supplies, (4) decrease in power consumption for oxygen production, (5) application of more efficient but complex heat recovery systems, for example, raising water vapor parameters in a heat-utilizing loop, or using a thermal design that includes an SG afterburner after the GTE.

5.2 Working Process in a Gas Turbine Combustor Operating on Synthesis Gas

Gas turbine engines are designed primarily to be fueled with natural gas, which mainly consists of methane. Recent increases in natural gas prices and concerns about its availability, as well as significant advances in synthesis gas production and cleanup, have made opportunities for the synthesis gas as a primary GTE fuel [15].

Plasma-assisted gasification can be efficiently used to convert carbon-containing materials to synthesis gas that can be used to generate power [6, 16–20].

Rather serious increase of toxic substances emission is possible when using synthetic fuels in GTE combustors in comparison to engines which operate on traditional fuel. Thus, there is a necessity to provide design solutions for mitigation of emission of hazardous elements.

Almost all the known methods of suppression of NO_x emissions (during synthetic fuel combustion) are connected with decreasing of temperature in the reacting zone or the volume of the high-temperature zones. These methods are the following: combustion of lean preliminary mixed air-fuel mixtures; combustion according to the "Rich-Burn, Quick-Mix, Lean-Burn" scheme (RQL scheme); water or water-steam injection to the combustor; catalytic burning [5, 21].

The RQL combustion scheme includes initial combustion of a rich air-fuel mixture, quick mixing and burning of a lean air-fuel mixture [22]. The combustor is divided into two zones. The first zone shall provide stable combustion of the fuel-rich mixture with a relatively low temperature and a low amount of an oxidant (nitrogen oxide formation is low under these conditions). The second zone (after quick and uniform mixing with an oxidant) provides combustion of the fuel-lean mixture at which nitrogen oxide formation is low as well. It should be mentioned that the main problem of this combustion technology is providing quick and qualitative mixing of the flows of an oxidant and the products of hydrocarbons partial oxidation at the intermediate stage (Quick-Mix) [23] in order to avoid forming a mixture with stoichiometric content in some local zones of the combustor. If there is stoichiometric content, there will be high temperatures and a high level of nitrogen oxide emission.

Selection of rational correlations of geometry parameters of this two-zone combustor which operates on synthesis gas and defining the most effective ways of oxidant (air) supply to the zone of quick mixing is a rather important practical issue. Its solution through experimental means requires substantial expenditures of material and human resources. Therefore, using the computational models will provide sharp reduction of time and resources for designing an RQL-combustor working on the synthesis gas.

The offered CFD mathematical model of a combustor working on the synthesis gas is based on the following equations [24–26]: continuity equation, momentum conservation equation, energy conservation equation, and mass conservation equation for main chemical components of the mixture. These equations are closed by adding proper differential equations of the turbulence model.

The Eddy Dissipation Concept (EDC) model is used for modeling of the synthesis gas combustion processes. It is an expanded model of turbulent eddy dissipation, which includes low-level modeling of chemical processes of fuel oxidation under the conditions of flame with turbulent flow excitations [27].

The studies of operation of modern gas turbine combustors show that the burners developed according to the principles of partial preliminary mixing of fuel with an oxidant have the tendency for non-stable behavior when operating with synthesis gas, which is the fuel with a high amount of CO and H_2, when diffusion-type combustors show more stable results.

If fuel has such components as CO, H_2 and CH_4, then in terms of numerical modeling of combustion processes there should be substantiated selection of the most proper kinetic schemes for defining the main flame parameters at reasonable computational efforts.

The detailed expanded chemical mechanisms which describe combustion of hydrocarbon fuel are developed properly for combustion of CO/H_2 mixtures. There are also additional simplified, or so-called global, mechanisms for CFD modeling which are mainly used for calculations of oxidation of hydrocarbon fuels and synthesis gases. For example, Bohni and others developed a six-stage kinetic mechanism by means of systematic decrease of the detailed kinetic mechanism [28].

Two mechanisms of chemical kinetics were used in this investigation to provide the detailed analysis of operating processes: (1) a simplified 35-reaction Reduced mechanism (Table 5.2) obtained by changing the multi-reaction kinetic scheme GRI-Mech [29], and (2) a kinetic Yetter mechanism, offered by Yetter R.A. (Pennsylvania State University), Dryer F.L., and Rabitz H. (Princeton University) [30] (Table 5.3).

The main chemical components of the Reduced mechanism are CH_4, O_2, CO_2, CO, H_2O, H, OH, O, H_2, HO_2, CH_3, HCO, CH_3O, CH_2O, H_2O_2, N_2.

The main chemical components of the Yetter mechanism are H_2, H, O_2, O, OH, H_2O, HO_2, H_2O_2, CO, CO_2, HCO, N_2, AR.

To verify the mathematical model of a combustor operating on synthesis gas, an ejection type afterburner of synthesis gas [20] was chosen as a research object.

For designing an afterburner, the following initial data were accepted: synthesis gas consumption of 25 g/s; temperature of 700 K and below; air flow via an afterburner of 400 g/s, air temperature of 350 K. As the pressure of the obtained synthesis gas is close to the atmospheric one, an original ejection system is developed for providing synthesis gas supply to the afterburner.

The data acquisition scheme is shown in Fig. 5.6 [20].

Air from the compressor 3 is fed into the afterburner full-size model 1 through the Coriolis flowmeter 2 with a capacity of 0–500 g/sec. Air temperature is measured using temperature sensors 4 with an operating range of 0 to 373 K, pressure—with manometer 5 with an operating range of 0 to 0.15 MPa. The "Comb" of thermocouples 6 with the operating temperature range of 573 to 1573 K and emission analyzer Testo 350 are mounted in the outlet nozzle. The data from them are displayed on a personal computer 8 via the digital converters 7. The pilot fuel gas required to maintain the combustion in afterburner is supplied from the high-pressure cylinder 9 via the pressure-reducing valve 10. Simulating gas flow is controlled by Coriolis

Table 5.2 Reactions of reduced mechanism

$H + O_2 \rightarrow OH + O$;	$OH + O \rightarrow H + O_2$;	$O + H_2 \rightarrow OH + H$;
$OH + H \rightarrow O + H_2$;	$OH + H_2 \rightarrow H_2O + H$;	$H_2O + H \rightarrow OH + H_2$;
$OH + OH \rightarrow H_2O + O$;	$H_2O + O \rightarrow OH + OH$;	$H + O_2 + M \rightarrow HO_2 + M$;
$HO_2 + H \rightarrow OH + OH$;	$HO_2 + H \rightarrow H_2 + O_2$;	$HO_2 + OH \rightarrow H_2O + O_2$;
$CO + OH \rightarrow CO_2 + H$;	$CO_2 + H \rightarrow CO + OH$;	$CH_4(+M) \rightarrow CH_3 + H(+M)$;
$CH_3 + H(+M) \rightarrow CH_4(+M)$;	$CH_4 + H \rightarrow CH_3 + H_2$;	$CH_3 + H_2 \rightarrow CH_4 + H$;
$CH_4 + OH \rightarrow CH_3 + H_2O$;	$CH_3 + H_2O \rightarrow CH_4 + OH$;	$CH_3 + O \rightarrow CH_2O + H$;
$CH_2O + H \rightarrow HCO + H_2$;	$CH_2O + OH \rightarrow HCO + H_2O$;	$HCO + H \rightarrow CO + H_2$;
$HCO + M \rightarrow CO + H+M$;	$CH_3 + O_2 \rightarrow CH_3O + O$;	$CH_3O + H \rightarrow CH_2O + H_2$;
$CH_3O + M \rightarrow CH_2O + H+M$;	$HO_2 + HO_2 \rightarrow H_2O_2 + O_2$;	$H_2O_2 + M \rightarrow OH + OH + M$;
$OH + OH + M \rightarrow H_2O_2 + M$;	$H_2O_2 + OH \rightarrow H_2O + HO_2$;	$H_2O + HO_2 \rightarrow H_2O_2 + OH$;
$H + OH + M \rightarrow H_2O + M$;	$H + H+M \rightarrow H_2 + M$;	

flowmeter 11 with a capacity of 0 to 30 g/s, while pressure is controlled by a vacuum gauge 12.

The operating processes in the afterburner were studied at atmospheric pressure (101325 Pa). The mass air consumption varied from 50 to 400 g/s. Propane-butane is used as a pilot gas.

The obtained data were verified by comparison of the main afterburner parameters (consumption of the ejected air and the amount of toxic components on the exhaust) obtained by computational and experimental methods.

The comparison of the calculated and experimental values of the afterburner parameters in the modes of cool and hot blow are shown in Figs. 5.7 and 5.8 [20].

The analysis of the obtained results shows the acceptable level of coincidences of the experimental and calculated values, which proves the adequacy of the offered mathematical and physical models and possibility of their use for calculating the gas turbine combustors which operate on synthesis gas.

To study the influence of the synthesis gas content on the gas turbine combustor parameters using a computer-aided design system, a parameter digital model of ½ part of a traditional diffusion-type combustor of the gas turbine engine UGT2500 [31] was designed (Fig. 5.9).

Table 5.3 Reactions of yetter mechanism

$H + O_2 = O + OH$	$H + O_2 + N_2 = HO_2 + N_2$	$H_2O_2 + O = OH + HO_2$
$O + H_2 = H + OH$	$H + O_2 + AR = HO_2 + AR$	$H_2O_2 + OH = H_2O + HO_2$
$OH + H_2 = H + H_2O$	$HO_2 + H = H_2 + O_2$	$CO + O + N_2 = CO_2 + N_2$
$OH + OH = O + H_2O$	$HO_2 + H = OH + OH$	$CO + O + AR = CO_2 + AR$
$H_2 + N_2 = H + H + N_2$	$HO_2 + O = OH + O_2$	$CO + O_2 = CO_2 + O$
$H_2 + AR = H + H + AR$	$HO_2 + OH = H_2O + O_2$	$CO + OH = CO_2 + H$
$O + O + N_2 = O_2 + N_2$	$HO_2 + HO_2 = H_2O_2 + O_2$	$CO + HO_2 = CO_2 + OH$
$O + O + AR = O_2 + AR$	$H_2O_2 + N_2 = OH + OH + N_2$	$HCO + N_2 = H + CO + N_2$
$O + H + M = OH + M$	$H_2O_2 + AR = 2OH + AR$	$HCO + H = CO + H_2$
$H + OH + N_2 = H_2O + N_2$	$H_2O_2 + H = H_2O + OH$	$HCO + O_2 = CO + H O_2$
$HCO + O = CO + OH$	$H_2O_2 + H = H_2 + HO_2$	$HCO + OH = CO + H_2O$
$H + OH + AR = H_2O + AR$	$HCO + AR = H + CO + AR$	

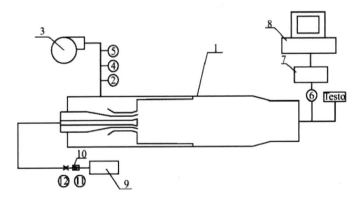

Fig. 5.6 Scheme of the measurements

The calculations were performed for three variants of fuel gas content. Variant 1 provided methane supply to the combustor, the results of its calculations were compared to the results of experimental studies of a traditional diffusion-type

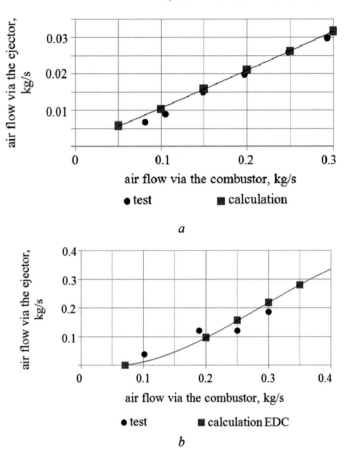

Fig. 5.7 Experimental and calculated dependences of air consumption via the ejector on air consumption via the combustor: a—cool blow mode; b—hot blow mode

Fig. 5.8 Experimental and calculated dependences of nitrogen oxides emission on air consumption via the afterburner

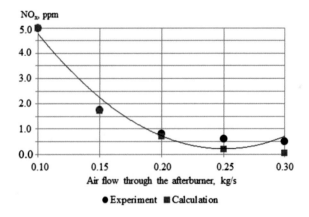

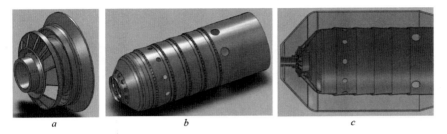

Fig. 5.9 Geometry model of the combustor: a—swirler; b—combustion liner; c—longitudinal section of the combustor

combustor and used for verification of the kinetic scheme of hydrocarbons oxidation. Variants 2 and 3 suggest supply of synthesis gas of various content and various calorific values to the combustor (Table 5.4).

To provide the same heat power of a burning device which corresponds to the nominal mode of a GTE, the synthesis gas flows were increased in comparison to the methane flow proportionally to reduction of the lower calorific value. The areas of the flow sections of a burning device were increased, which provided the required velocity of synthesis gas flow from the fuel holes without flame blow out. At this stage, apart from the change of the flow sections of a burning device, no design upgrade of the combustor was made.

Table 5.4 Initial data for gas turbine combustor calculation

	Variant 1	Variant 2	Variant 3
Lower calorific value, kJ/kg	48,800	21,791	12,448
Gas Content, % (vol.)			
CO_2	0	12.59	2.66
H_2O	0	0	4.88
CO	0	15.09	59.39
H_2	0	50.63	31.76
CH_4	100	19.08	0
N_2	0	2.61	1.31
O_2	0	0	0
Stoichiometric air amount, kg/kg	16.72	6.70	3.21
Air flow via combustor, kg/s	6.91		
Air temperatureon combustor inlet, K	650.0		
Air pressureon combustor inlet, MPa	11.91		
Fuel flow via combustor, kg/s	0.097	0.217	0.375
Fuel temperature, K	303		
Total air excess coefficient	4.26	4.74	5.74
Average fuel velocityin outflow holes, m/s	200	218	200

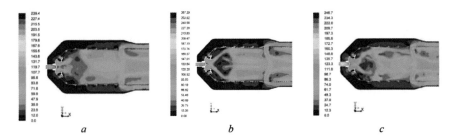

Fig. 5.10 Distribution of the velocity in the longitudinal section of the combustor, m/s: a—variant 1; b—variant 2; c—variant 3

As the result of calculations with the use of the EDC combustion model and the simplified Reduced kinetic mechanism [29], the following data on distribution of velocities, temperatures and concentrations of chemical components in the combustor were obtained.

Figure 5.10 shows the velocity distribution in the combustor. The specific growth of the flow velocity on the outlet of the swirler at reduction of the synthesis gas calorific value is connected with the corresponding growth of fuel flow.

For variant 3 the flow velocity in the area of the output section of the swirler reaches 120 m/s, while for the combustor operating on methane it is on the rational level of about 80 m/s. This feature shows the necessity of changing open flow areas in the swirler to provide the calculated velocities in the combustion zone during combustor modification.

It is possible to analyze the details of the changing character of the working medium velocities using Fig. 5.11, where the distribution of the axial velocity component in the longitudinal combustor section is shown. Variants 1 and 2 show the presence of the stable symmetry zone of reverse flows which stabilize combustion. In variant 3 (with the lowest calorific value of synthesis gas) the reverse flows zone is much more diluted, which determines lower stabilizing action of the combustion products. It should be mentioned that for variant 3 the asymmetry of the flows in

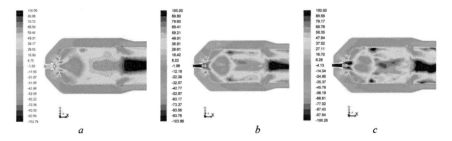

Fig. 5.11 Distribution of the axial velocity component in the longitudinal section, m/s: a—variant 1; b—variant 2; c—variant 3

the combustion liner starts appearing, which indirectly proves lower stability of the flame compared to variants 1 and 2.

Figures 5.12 and 5.13 show the diagrams of the dependences of average mass temperature along the combustor and the temperature contours in the combustion liner. As the fuel gas calorific value decreases, the combustion zone moves away from the burning device and shifts to the combustion liner outlet.

Asymmetry of the temperature field for variant 3 with the lowest calorific value of synthesis gas shows combustion non-stability, which requires development of measures to improve the efficiency of the flame front stabilization. The presence of methane in synthesis gas content (variant 2) stabilizes temperature distribution.

There is no decrease of the maximum temperature in the combustor volume in connection with decrease of the fuel gas calorific value, as the diffusion burning principle is implemented in the considered gas turbine combustor.

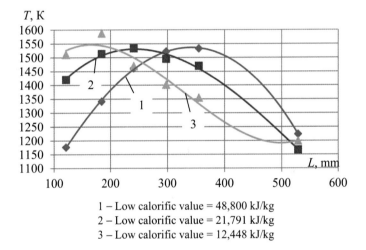

1 – Low calorific value = 48,800 kJ/kg
2 – Low calorific value = 21,791 kJ/kg
3 – Low calorific value = 12,448 kJ/kg

Fig. 5.12 Change of average mass temperature (T) along the length (L) of the combustor

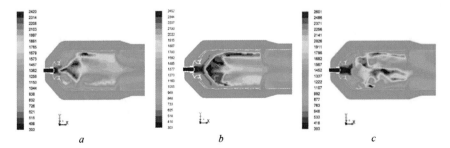

Fig. 5.13 Temperature contours in the longitudinal section of the combustor, K: a—variant 1; b—variant 2; c—variant 3

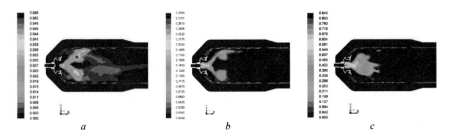

Fig. 5.14 Distribution of CO mass fractions in the longitudinal section of the combustor: a—variant 1; b—variant 2; c—variant 3

Distribution of carbon oxide CO in the combustion liner is shown in Fig. 5.14. The growth of the CO maximum concentration in the primary combustor zone from variant 1 to variant 3 is logical as the initial carbon oxide content in fuel gas for variant 3 is maximum. In general, in comparison to variant 1, the mole fraction of CO on the output decreased by 44.3% for variant 2 and by 50% for variant 3. Thus, CO content on the combustion liner outlet decreased simultaneously with decrease of methane content in fuel gas despite the fact that CO fraction in the initial synthesis gas grows.

Nevertheless, considering the obtained data, it is possible to stipulate that the growth of the combustion volume is positive for the level of CO burning-off and the combustion efficiency.

Figure 5.15 shows distribution of nitrogen oxides in the combustor. With the decrease of the synthesis gas calorific value and increase of the velocity of gases flow in the combustor, the zone of NO formation shifts to the outlet section. Thus, the NO emission for variant 3 with the minimum synthesis gas calorific value is

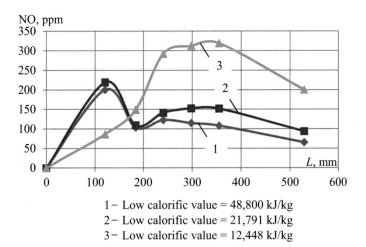

Fig. 5.15 Distribution of NO volume fractions in the longitudinal sections of the combustor

Table 5.5 Results of combustor calculations

	Variant 1	Variant 2	Variant 3
Low calorific value, kJ/kg	48,800	21,791	12,448
Temperature on combustor outlet, K	1224	1167	1199
Velocity on combustor outlet, m/s	74.18	74.56	81.35
Mole fraction of CH_4	0.00,042	0.00,022	0
Mole fraction of O_2	0.1544	0.1604	0.1631
Mole fraction of CO_2	0.0235	0.0261	0.0508
Mole fraction of H_2O	0.0501	0.051	0.0304
Content of N_2O, ppm	0.363	0.260	0.0393
Content of NO, ppm	65.7	94.0	200.3
Total pressure losses, %	4.51	4.31	4.03

about 200 ppm. Nitrogen oxides emission of the combustor operating on synthesis gas which contains methane (variant 2) is 94 ppm. The calculated emission of nitrogen oxides in the combustor operating on methane (variant 1) is 66 ppm.

It should be mentioned that air redistribution along the combustion liner at further combustor modification will allow avoiding the local values of the air excess coefficient at which nitrogen oxides emission is maximum and reducing the volume of the zones of possible nitrogen oxides formation.

The results of the performed three-dimension CFD calculations are shown in Table 5.5. The total pressure losses in the combustor were determined as well.

At calculating a gas turbine unit cycle, it was determined that the temperature on outlet of the combustor of the GTE with the power of 2.5 MW is 1224 K when operating on natural gas (that corresponds to the experimental data [31]), and the maximum temperature in the combustion zone is 2363 K. Thus, the applied mathematical model with the simplified Reduced mechanism [30] allows adequately predicting the temperature parameters of the combustor.

Unfortunately, selection of chemical mechanisms of synthesis gas combustion in GTE has been indefinite so far. Therefore, there has been performed an investigation of the influence of kinetic schemes on parameter distribution in the volume of the gas turbine combustor operating on synthesis gas with the minimum low calorific value (variant 3). Two detailed mechanisms of chemical kinetics were used for comparative calculations: the simplified 35-reaction Reduced mechanism [29] (mechanism 1) and the 70-reaction mechanism [30] (mechanism 2) developed for modeling of combustion of fuels which do not contain methane.

The results of modeling are shown in Table 5.6 and Figs. 5.16 and 5.17.

The results of the calculations show that there is no significant difference in contours of the velocities at using both mechanisms (Fig. 5.16). The temperature fields also do not have significant differences (Fig. 5.17). Distribution of the

Table 5.6 Calculation results for combustor using various chemical kinetics mechanisms

	Mechanism 1	Mechanism 2
Low calorific value, kJ/kg	12,448	12,448
Temperatureon combustor outlet, K	1199	1150
Velocityon combustor outlet, m/s	81.35	79.98
Mole fraction of CH_4	0	0
Mole fraction of O_2	0.1631	0.1665
Mole fraction of CO_2	0.0508	0.0457
Mole fraction of H_2O	0.0304	0.0276

Fig. 5.16 Distribution of the velocities in the longitudinal section of the combustor, m/s: a—mechanism 1; b—mechanism 2

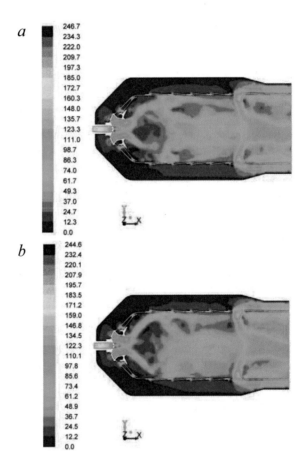

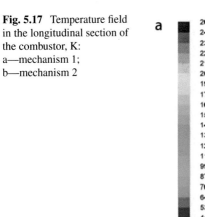

Fig. 5.17 Temperature field in the longitudinal section of the combustor, K:
a—mechanism 1;
b—mechanism 2

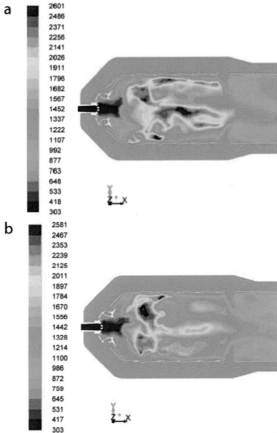

concentrations of all the chemical components of the oxidation scheme has similar nature.

Thus, we can recommend both the kinetic mechanisms for determining the main parameters of the combustor operating on synthesis gas.

It should be mentioned that emission parameters of the series production combustor of the GTE with the power of 2.5 MW due to its conversion from natural gas to synthesis gas without design changes of the main elements (for example, a combustion liner) are unsatisfying in terms of modern requirements to the toxic components emission. Thus, there should be development of the effective arrangements connected with the use of prospective schemes of the synthesis gas combustion organization.

5.3 Low-Emissive Combustion of Synthesis Gas

Numerical study of the 2.5 MW gas turbine combustor showed the necessity to provide significant changes to the scheme of the combustor working process organization in order to increase stability and efficiency of its operation, especially when operating on low calorific synthesis gas.

Simple replacement of natural gas (methane) with synthesis gas for a series-produced gas turbine combustor leads to increase of toxic components emission, particularly nitrogen oxides.

To improve parameters of the series-produced UGT2500 combustor [31], the numerical calculations of its modifications were performed, which were aimed at selection of rational geometry and mode parameters of a prospective two-zone combustor.

Initially, the following changes were made in the series-produced combustor design:

1. The amount of air supplied to the primary zone of combustion was reduced due to shifting from a barrier cooling system to a convective one and formation of an RQL-scheme of synthetic fuel combustion.
2. The air swirler was moved beyond the combustion liner and the form of the primary combustion zone was changed to the conic one.
3. Special plugs (shells) were added to the primary air holes to increase the depth of entering of air flows and improve the quality of components mixing on the output.

To reduce the possibility of formation of local areas in the combustor with the stoichiometric value of the air excess coefficient, reduction of air suctions to the primary zone (Rich-Burn) it is offered to pinch the section of the combustion liner before the holes of primary air supply.

For modified designs 1 and 2, the diameter of the corresponding shell ring of the combustion liner was reduced by 10% (Fig. 5.18a) and by 55% (Fig. 5.18b) in comparison with basic design. For design 2, the decision was made not to change the form and size of the combustor casing in order to minimize the structure changes of the basic construction. Higher pinching of the combustion liner section in this

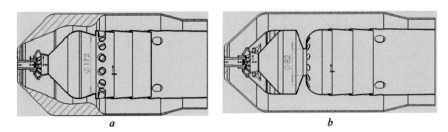

Fig. 5.18 Structure designs of combustors: a—modified design 1; b—modified design 2

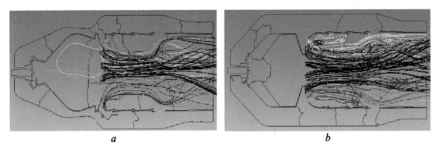

Fig. 5.19 Lines of the air tracks of the quick mixing (Quick-Mix)zone: a—modified design 1; b—modified design 2

case allowed avoiding installation of special plugs for increasing of primary air jets penetration.

The graph dependences show the minimum air suction to the side of the swirler in design 2 (Fig. 5.19b). It was favorable for reduction of the maximum temperature in the primary combustion zone which decreased from 2400 K to 2150 K for modified design 2 (Fig. 5.20b).

Despite this, there is no significant increase of carbon oxide emission in the output section of the combustion liner (Fig. 5.21). The change of the areas of the ring shells flow sections (at shifting from the rich air-fuel mixture combustion zone to the quick mixing zone) additionally provides slowdown of the flow and more qualitative mixing of air-fuel mixture with an additional oxidant.

The obvious effect of lack of primary air suction to the rich air-fuel mixture combustion zone is a sharp decrease of the calculated concentrations of nitrogen oxides in the outlet section to 29 ppm for modified design 2 (Fig. 5.22b, Table 5.7) which corresponds to modern environmental requirements. There are almost no local areas of thermal nitrogen oxides formation in the primary combustion zone for design

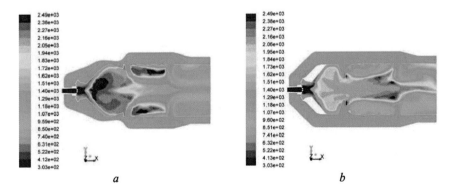

Fig. 5.20 Temperature field in the longitudinal section of the combustor, K: a—modified design 1; b—modified design 2

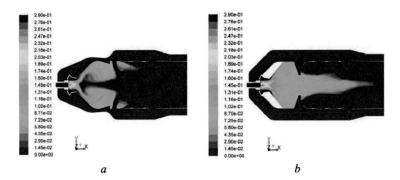

Fig. 5.21 Distribution of CO mass fractions in the longitudinal sections of the combustor: a—modified design 1; b—modified design 2

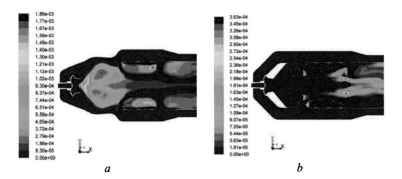

Fig. 5.22 Distribution of NO mass fractions in the longitudinal sections of the combustor: a—modified design 1; b—modified design 2

Table 5.7 Results of modeling of modified combustor designs operating on synthesis gas		Design 1	Design 2
	Temperature on the combustor outlet, K	1192	1190
	Velocity on combustor outlet, m/s	82.02	75.00
	Mole fraction of CH_4	0.000,001	0.00,001
	Mole fraction of O_2	0.1582	0.1577
	Mole fraction of CO_2	0.0277	0.0276
	Mole fraction of H_2O	0.0537	0.0544
	Content of N_2O, ppm	0.173	0.225
	Content of NO, ppm	109.1	28.9

2. This effect is possible due to provision of rational values of the air excess coefficients in the sections of the combustion liner. It was 0.54 in the primary zone and 2.4-2.6 in the secondary zone.

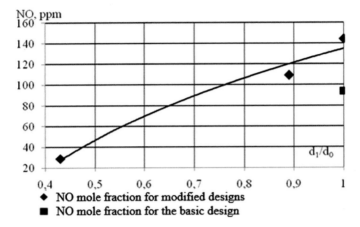

NO mole fraction for modified designs
NO mole fraction for the basic design

Fig. 5.23 Emission of nitrogen oxides NO in the combustor operating on synthesis gas

The given calculations showed that organization of the working process in the gas turbine combustor operating on synthesis gas is effective provided that the ratio of the diameter of the reduced ring shell part (d_1) to the diameter of the cylinder part (d_2) of the combustion liner is 0.40–0.55 (Fig. 5.23).

Since at operation of the gas turbine combustor on synthetic fuel (to provide its constant heat capacity) the synthesis gas flow grows to a high rate in comparison to the natural gas flow, there is practical interest for the interval of the values of the synthesis gas calorific value which allows implementing its stable and effective burning off without additional supply of natural gas.

The given variant calculations assumed supply of synthesis gas of various content and calorific value obtained by plasma treatment of low-grade coal [16] to the combustor of modified design 2: case 1—lower calorific value of synthesis gas of 33,089 kJ/kg, case 2—21,791 kJ/kg, case 3—12,448 kJ/kg.

The data of three-dimension CFD calculations (Table 5.8 and Fig. 5.24) performed using the simplified Reduced mechanism show that when a combustor of the 2.5 MW GTE operates on synthesis fuel, the minimum value of the synthesis gas low calorific value which provides its effective burning off without additional supply of natural gas is about 20 MJ/kg.

Table 5.8 Calculation results of the combustor operating on various-content synthesis gas

	Case 1	Case 2	Case 3
Low calorific value, kJ/kg	33,089	21,791	12,448
Temperature on the combustor outlet, K	1182	1191	1300
NO content on the outlet, ppm	31.4	28.9	116.4
Total pressure losses, %	4.72	4.71	3.87

Fig. 5.24 Combustor characteristics: a—change of the nitrogen oxides emission; b—change of average temperature along the combustor operating on various-content synthesis gas

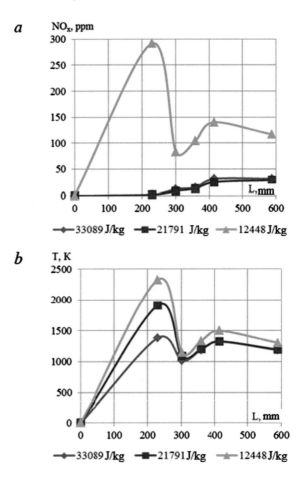

Increase of the values of the maximum temperature in the combustion zone (Fig. 5.24b) and the temperature on the combustor outlet for synthesis gas with the minimum low calorific value (case 3) is explained by significant increase of the fuel gas flow rate, its outflow velocity from the burning device, and the change of the aerodynamic structure of the reacting flows, which leads to worsening of the combustor ecological parameters.

5.4 Conclusions

Based on the results of the investigations, the following conclusions can be formulated.

1. Numerical study showed the necessity to provide significant changes to the scheme of the combustor working process organization in order to increase

stability and efficiency of its operation, especially when operating on low calorific synthesis gas. The use of the promising RQL combustion technology is proposed to improve the characteristics of the GTE combustor operating on synthesis gas.

2. The three-dimensional mathematical model and different kinetic mechanisms of a synthesis gas combustion processing have been used for enhancement of the 2.5 MW gas turbine ecological parameters.

3. There are developed practical recommendations which can reduce the nitrogen oxides emission up to 29 ppm in the exit section of the gas turbine combustors working on synthesis gas.

References

1. I. B. Matveev, S. I. Serbin, S. M. Lux, "Efficiency of a hybrid-type plasma-assisted fuel reformation system", IEEE Trans. Plasma Sci., vol. 36, no. 6, pp. 2940–2946, 2008.

2. S. I. Serbin, I. B. Matveev, "Theoretical Investigations of the Working Processes in a Plasma Coal Gasification System", IEEE Trans. Plasma Sci., vol. 38, no. 12, pp. 3300–3305, 2010.

3. S. I. Serbin, "Modeling and Experimental Study of Operation Process in a Gas Turbine Combustor with a Plasma-Chemical Element", Combustion Sci. and Technology, vol. 139, pp. 137–158, 1998.

4. S. I. Serbin, "Features of liquid-fuel plasma-chemical gasification for diesel engines," IEEE Trans. Plasma Sci., vol. 34, no. 6, pp. 2488–2496, 2006.

5. S. I. Serbin, I. B. Matveev, G. B. Mostipanenko, "Investigations of the Working Process in a "Lean-Burn" Gas Turbine Combustor With Plasma Assistance", IEEE Trans. Plasma Sci., vol. 39, no. 12, pp. 3331–3335, 2011.

6. I. Matveev, S. Serbin, "Investigation of a reverse-vortex plasma assisted combustion system", Proc. of the ASME 2012 Summer Heat Transfer Conference, July 8-12, Puerto Rico, USA, HT2012-58037, pp. 1–8, 2012.

7. Yu. D. Korolev, O. B. Frants, N. L. Landl, V. G. Geyman, I. A. Shemyakin,A. A. Enenko, I. B. Matveev, "Plasma-Assisted Combustion System Based on Nonsteady-State Gas-Discharge Plasma Torch", IEEE Trans. Plasma Sci., vol. 37, no. 12, pp. 2314–2320, 2009.

8. Igor B. Matveev, "Plasma or Retirement. Alternatives to the Coal-Fired Power Plants," IEEE Trans. Plasma Sci., vol. 39, no. 12, pp. 3259–3262, 2011.

9. I. Matveev, N. Washchilenko, S. Serbin, N. Goncharova, "Combined Cycle Gas Turbine Power Plant with Integrated Plasma Coal Gasification", in Proc. 8th Int. Conference on Plasma Assisted Technologies, Rio de Janeiro, Brazil, pp. 193–196, 2013.

10. I. Matveev, V.E. Messerle, A.B. Ustimenko, "Plasma Gasification of Coal in Different Oxidants," IEEE Trans. Plasma Sci., vol.36, no.6, pp. 2947–2954, 2008.

11. I. Matveev, S. Matveeva, "Some Results of Development and Experimental Investigations of High Power and High Pressure RF Torches with Reverse Vortex Plasma Stabilization", in Proc. 8th Int. Conference on Plasma Assisted Technologies, Rio de Janeiro, Brazil, pp. 55–57, 2013.

12. B. G. Trusov, "Program System TERRA for Modeling of Phase and Chemical Equilibrium", Int. Conference on chemical thermodynamics, St. Petersburg, 2002.

13. O. A. Lavrichshev, V. E. Messerle, T. F. Osadchaya, A. B. Ustimenko, "Plasma gasification of coal and petrocoke", 35th EPS Conference on Plasma Phys. Hersonissos, 9–13 June 2008, ECA Vol.32D, O-2.018, 2008.

14. G. F. Romanovsky, S. I. Serbin, V. N. Patlajchuk, Modern Gas Turbine Units, Mikolayiv: National University of Shipbuilding, 344 p., 2005.

15. S. Gadde, J. Xia, and G. McQuiggan, "Advanced F class gas turbines can be a reliable choice for IGCC applications", Siemens Power Generation, Inc., 2006.

16. I. B. Matveev, N.V. Washcilenko, S.I. Serbin, and N. A. Goncharova, "Integrated Plasma Coal Gasification Power Plant", IEEE Trans. Plasma Sci., vol. 41, no. 12, pp. 3195–3200, 2013.
17. I. B. Matveev, S. I. Serbin, and N. V. Washchilenko, "Sewage-to-Power", IEEE Trans. Plasma Sci., vol. 42, no. 12, pp. 3876–3880, Dec. 2014.
18. S. I. Serbin, I. B. Matveev, and N. A. Goncharova, "Plasma Assisted Reforming of Natural Gas for GTL. Part I", IEEE Trans. Plasma Sci., vol. 42, no. 12, pp. 3896–3900, 2014.
19. I. B. Matveev, N. V. Washchilenko, and S. I. Serbin, "Plasma-Assisted Reforming of Natural Gas for GTL: Part III - Gas Turbine Integrated GTL", IEEE Trans. Plasma Sci.,vol. 43, no. 12, pp. 3969–3973, 2015.
20. I. B. Matveev, S. I. Serbin, V. V. Vilkul, and N. A. Goncharova, "Synthesis Gas Afterburner Based on an Injector Type Plasma-Assisted Combustion System", IEEE Trans. Plasma Sci., vol. 43, no. 12, pp. 3974–3978, 2015.
21. H. J. Jung and E. R. Becker, "Emission control for gas turbines", Platinum Metals Rev., vol. 31 (4), pp. 162–170, 1987.
22. A. H. Lefebvre and D. R. Ballal, Gas Turbine Combustion: Alternative Fuels and Emissions, Taylor & Francis, 2010.
23. Rich burn, quick-mix, lean burn (RQL) combustor, http://www.netl.doe.gov/File%20Library/ Research/Coal/energy%20systems/turbines/handbook/3-2-1-3.pdf.
24. Michael Welch and Brian Igoe, "An introduction to, fuels, emissions, fuel contamination and storage for industrial gas turbines", Proceedings of ASME Turbo Expo 2015: Turbine Technical Conference and Exposition GT2015, June 15—19, 2015.
25. P. E. Rokke, J. E. Hustad, N. A. Rokke, and O. B. Svendsgaard, "Technology update on gas turbine dual fuel, dry low emission combustion systems", ASME, 2003.
26. J. D. Willis and A. J. Moran, "Industrial RB211 DLE Gas Turbine Combustion Update", ASME, 2000.
27. The Generalized Finite-Rate Formulation for Reaction Modeling, http://www.arc.vt.edu/ ansys_help/flu_th/flu_th_sect_finite_rate.html.
28. Andrea de Pascale, Marco Fussi, and Antonio Peretto, "Numerical simulation of biomass derived syngas combustion in a swirl flame combustor", ASME, GT-2010-22791, 2010.
29. An optimized detailed chemical reaction mechanism for methane combustion GRI-Mech 3.0, http://www.me.berkeley.edu/gri_mech/.
30. R. A. Yetter, F. L. Dryer, and H. Rabitz, "A comprehensive reaction mechanism for carbon monoxide/hydrogen/oxygen kinetics", Combust. Sci. Tech., vol. 9, pp. 97–128, 1991.
31. Gas turbine power plants, http://eng.zorya.com.ua/files/energi_eng.pdf.

Printed in the United States
by Baker & Taylor Publisher Services